SIGNALS AND COMMUNICATION TECHNOLOGY

For further volumes:
http://www.springer.com/series/4748

Apurba Das

Digital Communication

Principles and System Modelling

Springer

Apurba Das
Scientist, Image Processing Lab
CDAC, Kolkata
A Scientific Society
Ministry of Communication and IT
Government of India
West Bengal
India
apurba.das@cdackolkata.in
apurba_das1@hotmail.com

Additional material to this book can be downloaded from http://extra.springer.com

ISBN 978-3-642-12742-7 e-ISBN 978-3-642-12743-4
DOI 10.1007/978-3-642-12743-4
Springer Heidelberg Dordrecht London New York

Library of Congress Control Number: 2010929759

Cover design: WMXDesign GmbH, Heidelberg

Printed on acid-free paper

Springer is part of Springer Science+Business Media (www.springer.com)

To You
Angana (Oli),
my wife

Preface

In today's world of human-machine interaction, the necessity of custom communication is increasing day by day. The adjective 'custom' in communication or in true sense, telecommunication signifies a lot of things like secured communication, communication with hidden data through any type of multimedia covers like image, speech or video, communication in user suggested bandwidth, multi-channel communication without cross-talk, internetworking in LAN, MAN, WAN and integrated service digital network (ISDN) of image, video and data, data transmission in encrypted but non-perceptible mode and so on. The information exchange between the source and the destination is generally done through modulated signal both in analog and digital form. Digital design and processing can not only be easily realizable, but also digital signal processing can help one designer to meet the practical needs of this era as discussed before.

Therefore, the digital communication has become an important subject of study for students of electronics and communication, computer science and information technology both in undergraduate and post-graduate levels.

The aim of the book is to represent the theory and application of the design philosophy of the subject Digital Communication systems in a unique but lucid form. There are resource books that are supreme pedagogical documents of this subject. It is a bit difficult to consider them as the standard text for the entire student population, because the said books do not visualize the design problem clearly. An attempt has been made to bridge the gap between the design principle and system modelling. In this book, I have tried to introduce the subject through its obvious flow. Supporting MATLAB codes and outputs are also included for better understanding and visualization. Essentially it is designed for the large class of students in the standards of bachelors, masters or those who have started their research carrier in science, engineering or technology.

The title of the proposed book is 'Digital Communication- Principles and system modelling'. By selecting this name I essentially took the task to insert equal importance to the theory and application aspect of the subject. The subject is introduced considering absolutely zero prerequisite of the readers. The introductory chapter created a space for round table discussion between the author and the reader in the subject of signals, systems, types and choices of telecommunication methodology. Basic building blocks play their roles to design one complete digital communication system according to necessity of successful interaction of information. In second chapter, the process of conversion of natural (analog) signal to digital signal is discussed for letting the input acceptable to digital communication system. In the subsequent chapters the processes of signal transmission with and without modulation are described. Process of security embedding in terms of spreading the bandwidth of a signal is represented. The issues of inter-symbol interference (ISI) are solved by using equalizer. Theory of information and channel capacity are also discussed in the light of probability theory. As the book claims zero prerequisite, one Appendix titled as *Elementary Probability Theory* is also appended to the main chapters. The error detection and correction in one digital communication system is understood physically and then suitable coding schemes are being employed to solve particular types of bit errors due to channel noise.

In the application part I have put my best effort to select a wide class of problems. Some of them are referral in nature and found in most of the books but the other group includes the applications, which I have collected from different resource books and different research papers (all the time submitting courtesy to the authors). Some applications are also represented as case studies on frequently used concepts convolution and correlation in Appendix B.

The style of writing is kept in the lucid level to attract the interest of a large class of readers. I always believe that, writing a text book is very much difficult in terms of presentation. Text book is not only for concept sharing, but also for preparing a slide show or movie which can just put the readers into a smooth and comfortable path of understanding the subject. Similarly, teaching is not a delivery of some raw data; conversely, it is a performance in front of the students/ audience. I think this book can be a good teaching aid, too.

Salient Features

1. The application area is rich and resemblance to the present trend of research.
2. An online content is included with the title, which includes *codes* and *MATLAB*, with illuminating uncommon and common applications along with some *flash movies* for better understanding of the subject.
3. Elegant worked out exercise section is designed in such a way that, the readers can get the flavour of the subject and get attracted towards the future scopes of the subject.
4. Unparallel tabular, flow chart based and pictorial methodology description is included for sustained impression of the proposed design/algorithms in mind.

The book is for everybody, the reader may be someone without the knowledge of basic engineering or someone like a researcher who wants to understand the subject in a different angle. Let me use the technical terminology to let you understand. Ultimately the received signal to you transmitted by this book must be converted successfully to analog/ natural form; otherwise the knowledge cycle does not get completed.

Kolkata, India Apurba Das

Acknowledgements

Human being is not at all an island. A person can be successful only when he/she is blessed and encircled by the relatives and community. This book would not be completed without the blessings of my seniors, the well wishes of my friends, colleagues and students.

First of all I like to express my deep respect to my parents Sri Barun Das and Smt. Nabanita Das for setting up a vision for me. I like to express my regards to my uncle and aunt (Jethu and Mamoni) Sri Arun Kr. Das and Smt. Moli Das for their continuous support. I am also thankful to my brother Anirban and sister Adwitia as they considered me as their benchmark of living.

I am fortunate enough getting Prof. Amit Konar of Electronics and Tele-Communication Engineering department of Jadavpur University as my teacher during my M-Tech. Sir has always inspired me about writing text books. He himself is the inspiration for thirst of knowledge.

I am grateful to my friend Mr. Manoj Kr. Pain for everything. Not only he helped me in preparing the flash files for online content, moreover he is the man behind all of my actions particularly in this phase of my life.

I thank my students, especially Dipto Pandit, Subarno Shekhar Gain, Subhojit Chatterjee for their help in content preparation for error control coding chapter and some parts of Delta Modulation.

I am really grateful to my senior Mr. Debasis Mazumdar of CDAC, Kolkata. I have learnt how nicely mathematical models can infer a physics and how a physics can generate one nice mathematics. I am also thankful to my colleague Mr. Raja Gupta for preparing the flash files for online content.

Last but not the least I like to express my thanks to the editorial team of Springer-Verlag, specially Christoph Baumann and Carmen Wolf for such a wonderful production.

Kolkata, India — Apurba Das

Contents

Chapter 1
Preview and Introduction

1.1 Process of Communication

A man is discussing about a cricket game with his friend sitting just in front of him. If this is the scenario, this process of interaction is interpersonal communication. In this case, no help of electronics is needed for communication. Let's now change the scenario. Say, a political meeting is going on at a large play ground. In front of 1,000,000 people one speaker is giving his/her lecture. It's not easy to communicate now, as because (i) the distance between the communicating nodes has been increased and (ii) the audience is not deaf and dumb therefore all of them are responsible for generating unwanted signal (noise). Here we require some extra circuits for successful communication. First the speech signal has to be converted to electrical equivalent signal known as audio by a transducer namely, microphone. Next, the electrical signal should be amplified by an audio amplifier. Finally, that amplified audio signal is re-converted by another transducer namely loud speaker to speech signal just as the amplified source signal to make it audible. The block diagram representation is shown in Fig. 1.1.

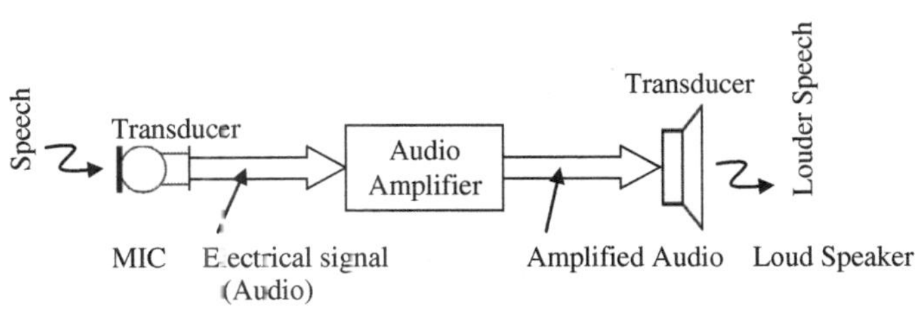

Fig. 1.1 Speech signal amplification with the help of transducers and electrical amplifier

Therefore we have started using the knowledge of electrical and electronics engineering in communication. Now, let's make the distance longer. Communication is required between two persons; one in USA is trying to communicate other persons in Germany. Here, microphone and loud speaker cannot also solve the present problem. Here two ways are there.

A. Das, *Digital Communication*, Signals and Communication Technology,
DOI 10.1007/978-3-642-12743-4_1, © Springer-Verlag Berlin Heidelberg 2010

1. There must be a long connecting cable namely transmission line between the persons. If this is the mode of communication, then that is known as ***Line Communication.***
2. Otherwise, we can take the help of an antenna to transmit and receive the un-modulated (base-band) or modulated audio signal, this is known as ***Radio wave communication.***

Combining the above two, the process of long distance communication is named as Tele-Communication. The meaning of the word 'tele' means far. Obviously, we are now convinced that, electronics must be needed for Tele-Communication. Our study through out the title should be based on Tele-Communication only.

A communication system can also be classified into three categories in respect of direction of communication [1] as:

1. Simplex
2. Half Duplex
3. Full Duplex

Simplex communication is a unidirectional communication system i.e., communication is possible in one direction only. Typically, the transmitter (the one talking or sending information in any ways) sends a signal and it's up to the other receiving device (the listener) to figure out what was sent. This type of communication is most efficient when there is a lot of data flow in one direction, and no traffic is needed in the other direction. Broadcast systems like the T.V and radio signals, fire alarm systems etc. are good examples of Simplex communication system.

In Half Duplex communication system, bi-directional communication is possible, but only in one direction at a time. That means one can either transmit or receive a signal at a particular instant of time in this system of communication. One cannot transmit and receive a signal simultaneously. Say, within a specific time interval $t_1 \leq t < t_2$, node A is transmitting a signal to another node B. At time $t = t_2$, node A will command 'over' and both of the nodes A and B will change their mode from transmitter to receiver and receiver to transmitter, respectively by switching the duplexer. Next, B can start sending information to A via the same link. The walky-talky used in defense and by police is a good example of Half Duplex communication system.

In Full Duplex communication system simultaneous two-way communication is achieved. Unlike half duplex communication system, one can both transmit and receive a signal simultaneously. Telephone conversation is an appropriate example of Full Duplex communication system (Fig. 1.2).

When single transmitter is sending the same information to a plenty of receivers, the special mode is called as broadcasting.

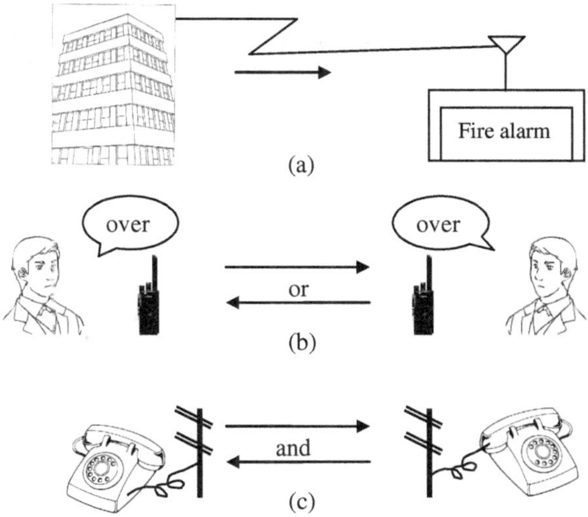

Fig. 1.2 Modes of directional communication: **(a)** Simplex, **(b)** Half-duplex, **(c)** Full-duplex

1.2 General Definition of Signal

In a single phrase, signal is the physical quantity by which information is carried between at least two entities. In other words by which physical quantity communication is done, is signal. Signal is not necessarily always represented as some electrical form of energy, it may be some gait, may be an image, may be a video.

To construct a complete definition let us try to have a close look at the general properties of a signal.

Property 1 Representation of a physical quantity.

Property 2 Signal must carry some information. Meaning of a signal is not important in the property discussion. If a signal does not have meaning with respect to a particular communication, it may be in coded form or meaningful for another arena of communication. It means the physical quantity is unwanted with respect to that specific application. Therefore, this is unwanted signal or noise.

Property 3 To measure or predict the response of a physical phenomena or system, mathematical modeling is important. Electrical current signal is generally characterized as

$$i(t) = I_0 \cos(\omega_0 t + \phi) \tag{1.1}$$

If we observe a small time window of that signal, the waveform obtained would be as shown in Fig. 1.3a.

Referring the Fig. 1.3, the time domain and frequency domain representations are shown. Obviously, the spectrum of the single tone signal is showing an impulse

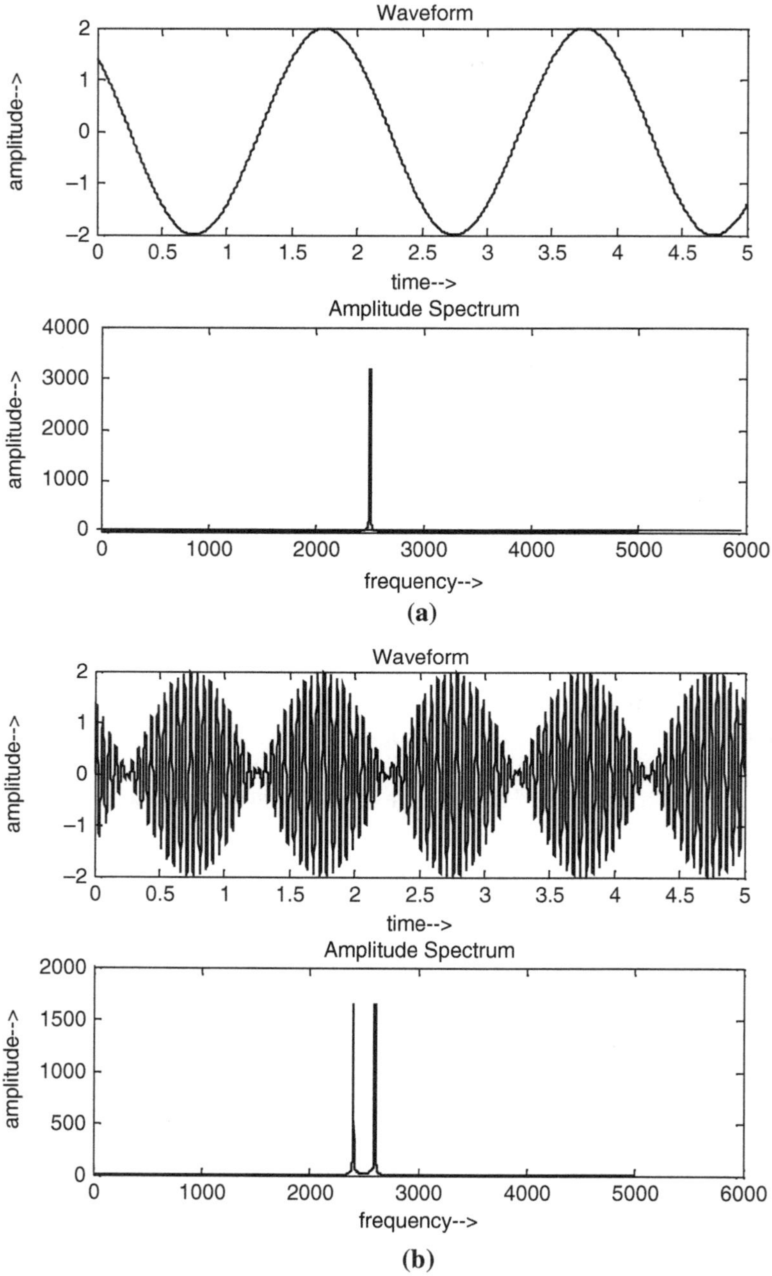

Fig. 1.3 **(a)** Time and frequency representation of a single tone signal, **(b)** Time and frequency domain representation of a two tone signal

only, whereas the two tone (DSB-SC type) signal generates a pair of impulses. These representations of math models are satisfying the definition of signal but can only be considered as an example of a special class of signal, which is modeled as function of time or frequency as f(t) or $F(\omega)$. Let's consider now an image shown in the following Fig. 1.4. To represent the image as a mathematical function, we have to first divide the entire image into a number of unit squares of homogeneous intensity value to each square. These smallest picture elements are called as pixels. Intensity or color of each smallest square block is now can be represented as function of ordinate and abscissa, i.e., function of space where the entire image plane is considered as a Cartesian co-ordinate system.

Fig. 1.4 Image representation as a function of 2D space

Similarly, video images can be represented functionally as a multivariate function, where image signal is represented as I(t,x,y), i.e., function of both time and space.

As per the discussion above, now we can characterize a signal mathematically as a function of independent variable/variables.

Now from the above three properties, we can define signal as, *A physical quantity used for carrying information from one entity to another and mathematically modeled as a function of independent variable/ variables.*

Our discussion would be mainly focused on the signals which can be modeled as a function of time. In the next section, we'll discuss the time-value definition and classification of signals. Hence, we'll clearly understand the time-value concept of analog and digital signals. Also we can find the step-by-step procedure of conversion of analog signal to digital signal.

1.3 Time-Value Definition of Signals–Analog and Digital

The word 'analog' comes from the word 'analogy' which means similarity. Here the similarity implies similarity with the nature. Directly we can say, any natural signal is analog signal. As nature, the starting time of analog signal is also $t = -\infty$ and the ending time is $t = \infty$.

Let's consider the waveform as shown in the Fig. 1.5. We have taken one time window of the signal though the existence of the signal is beyond the time range $t = [t_1\ t_2]$. Within this time range, the signal $x(t)$ has maximum and minimum values X_{max} and X_{min} respectively. Now depending on the countability of the number of possible amplitude and possible time values, we can classify signal and this is in turn the step by step approach towards the procedure of ADC (analog to digital conversion).

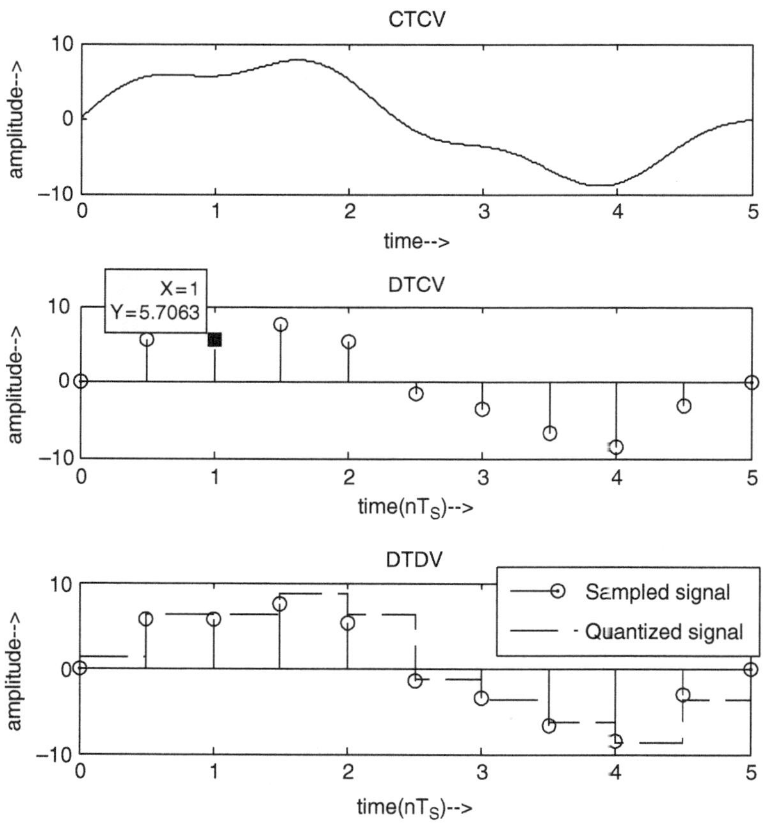

Fig. 1.5 Time valued definition of signal

1.3.1 Continuous Time Continuous Valued Signal

As shown in the Fig. 1.5, the signal $x(t)$ is considered within the time interval $t = [t_1\, t_2]$, say within time 10:10 a.m. and 10:11 a.m. This technique of watching a signal in a specific interval is called as windowing. This is easily understandable that, we can take any time value within the specified interval and any amplitude can also be obtained at any specific time instant t_i, i.e., the number of possible valid time instants is infinite. From the concept of set theory, it is considered that, from a set of finite number of time values, any valid time instant t_i must be the element of the predefined set. In the present case we cannot define any set of finite number of elements. Because, we can have any instant like 10 h 10 min 1 s 1.3 ns, so possible time values can be infinite. In the case of amplitude value also we can have infinite number of possible amplitude values within the range X_{max} and X_{min}. This type of signal is therefore called as continuous time continuous valued (CTCV) signal.

1.3.2 Discrete Time Continuous Valued Signal

Generally, the analog signals do have this property of continuous time and amplitude values. But no digital memory or processor can deal with infinite long words. Digital signals are represented using digits only. First, to express the CTCV signal, let's take the help of the equally samples of the signal. The equal time spacing between two adjacent samples is called as sampling interval, T_s. Then the 2nd signal shown in the figure must have finite number of time values but, we can have infinite number of possible voltages values yet. This signal is called as discrete time continuous valued signal (DTCV).

1.3.3 Discrete Time Discrete Valued Signal

As shown in Fig. 1.5, at t=1, the amplitude of the signal is 5.7063. It may be anything between X_{max} and X_{min}. Now, to get finite number of voltage values, a set of finite number of amplitude values setting the X_{max} and X_{min} as upper and lower limit, has to be defined. In the present case, within the amplitude range 10(X_{max}) and -10(X_{min}), 8 values have been taken to form that closed set. The elements are, ± 1.25, ± 3.75, ± 6.25, ± 8.75. Now, sampled ales can be approximated (i.e., quantized) to the closest defined element of the amplitude. As shown in Fig. 1.5, at time instant t=1, the sampled value 5.7063 has been quantized to 6.25. The staircase form of the signal is having finite number of time and amplitude values. This signal is called as discrete time discrete valued (DTDV) signal. Now, that defined amplitude values can easily be encoded to get the digital bit stream. The entire process is elaborated in the next chapter (Chap. 2).

1.4 Analog and Digital Communication Systems

In the process of telecommunication, one may be interested to transmit directly the natural signal (analog signal) with or without modulation (base band transmission). There, all the processors/systems used for transmission process are analog processors designed using discrete analog elements like op-amps, transistors, resistors, inductors, capacitors, etc. This is analog communication system.

On the other hand, the system which specifically deals with digital data and digitally pre-processed signals is a digital communication system. Here, digital processors and accessories like stored program controlled processors, software codes, digital memory etc. takes important role in the process of signal transmission and reception. But, as we know, all the natural understandable signals are analog, overhead hardware for analog to digital conversion and digital to analog conversion are needed at the transmitter and receiver end, respectively.

Although, the necessity of extra blocks for analog to digital conversion (ADC) and digital to analog conversion (DAC), digital communication system is preferred over analog communication system for some particular reasons as follows:

1. *Noise immunity*: While transmitting digital information, formatting is very essential, i.e., two analog correspondence amplitude values are to be assigned for logic zero and logic one respectively in case of binary transmission. Noise immunity into such formatted digital information is greater than that of an analog signal. An Additive White Gaussian Noise (AWGN), (additive in nature, constant PSD (Power Spectral Density) throughout all frequency components and Gaussian PDF (Probability Density Function)) is added to both analog and digital signals. From the Figure 1.6, it's clearly seen that the actual analog signal after being attacked by the noise is extremely distorted. Similarly the actual digital signal is also distorted but in case of digital signal the original signal can be easily reconstructed by simply taking the average amplitude at required time intervals. In the above example, the average value within time intervals (bit width) T_B, $2T_B$, $3T_B$ & $4T_B$ are somewhat nearer to +1 V, +1 V, 0 V & +1 V. Hence the original signal can be easily reconstructed.
2. *Memory:* Analog signals are generally stored in devices like magnetic tapes, floppy disks, etc. It requires many magnetic tapes to store the analog signals. Moreover these are easily affected by the magnetic and other mechanical and physical phenomenon. On the other hand, digital information is stored in devices like CDs and registers. For example, in a D-flip-flop, the output can be retained for many years without any external power. When the output is needed, the trigger clock pulse is supplied to the flip-flop & it will give the output. Lifetime of digital memory is also higher than that of analog memory.
3. *System Re-configurability:* One of the most significant advantages of digital systems is their ease of system re-configurability. As for an example, let us consider an analog low pass filter. To convert it to a high pass filter, we will have to remove the components from the circuit and replace them with other appropriate components. This is a tiresome job if the system is a complex one. On the other

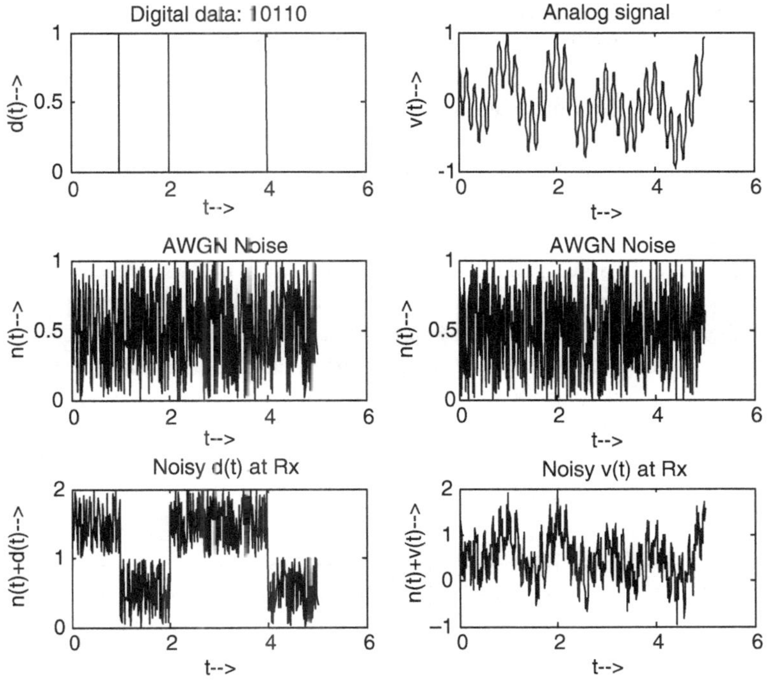

Fig. 1.6 AWGN attacked digital and analog signal

hand of we want to convert a digital low pass filter into a high pass filter, we can easily change the digital filter transfer function [H(z)] just by changing some coefficients ($a_0, a_1, a_2 \ldots, b_0, b_1, b_2, \ldots$).
Where

$$H(z) = \frac{a_0 + a_1 z^{-1} + a_2 z^{-2} + \ldots}{b_0 + b_1 z^{-1} + b_2 z^{-2} + \ldots}$$

Just taking the example of any multimedia audio player, we can get a clear view of the system. When the graphic equalizer is set to a 'high treble' preset, the entire system of reconstruction process responds as a highpass filter. Just at the very next instance if anyone chose to listen 'high bass' music, the reconstruction system responds like a lowpass filter. It's done only by dynamically change of coefficients values.

4. *Aging:* Aging signifies growing older of the system. It is obviously less effective in case of digital systems. In case of analog systems, the output may change after a few years due to aging of the discrete analog components like diode. As time goes, the cut-in voltage of the diode increases slowly and unsteadily. This causes fluctuation in the system performance. In digital systems, the system error due to the problem of aging is totally absent.

1.5 Elements of Digital Communication System

The system which specifically deals with digital data and digitally pre-processed signals is a digital communication system. In the Fig. 1.7, the basic elements of a digital communication system is shown. As we know, objective of transmitting natural information in terms of digital data from one end to another can be achieved by incorporating ADC at the transmitter side and DAC at the receiver side. Here, the message generated by the information source is to be converted to binary bit stream, first. At the time of this conversion, one should try to represent the message through as few binary digits as possible, i.e., an efficient information representation of the source message has been obtained, which results in very low or zero redundancy. This process of efficiently converting analog or digital message generated from source, into a sequence of binary digits is called as *source encoding*. This process includes digital *data compression*.

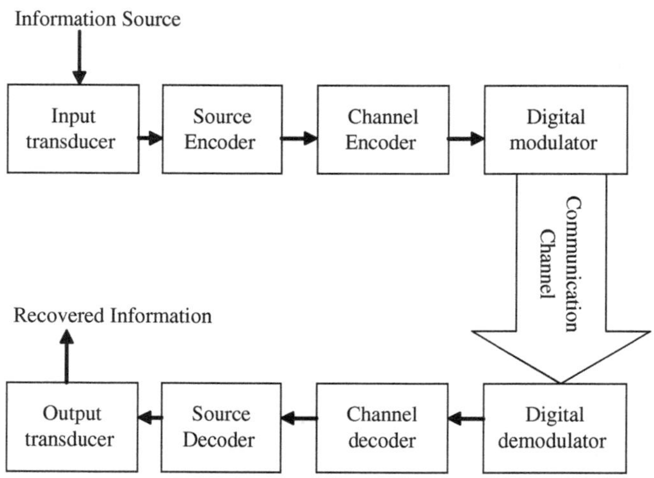

Fig. 1.7 Basic elements of a Digital Communication system

The sequence of bits is then passed through the channel encoder. The purpose of channel encoding is to introduce intentional custom redundancy into the information sequence to overcome the effects of noise encountered in the transmission channel at the time of communication. Let's consider a (trivial) form of channel encoding where each bit of the information sequence is simply repeated by m times. A better encoding (non-trivial) takes k information bits at the time of mapping each k-bit sequence into a unique n-bit sequence, called a *code word*[2]. The amount of redundancy introduced by encoding the data in this manner is measured by the ratio n/k. the reciprocal of the ratio k/n is called the *code rate*.

The channel encoded bits are then passed to a digital modulator. Digital modulation here signifies assigning an equivalent set of wave forms for channel encoded binary digits. To understand in brief, we can set $S_0(t)$ waveform is transmitted to transmit logic 0 and $S_1(t)$ waveform is transmitted to transmit logic 1. This technique of modulation can also be called as *binary modulation.*

Communication channel is the physical medium that is used to communicate the digital modulated/un-modulated signal from transmitter to the receiver end. In wireless communication, free space is considered as the channel. On the other hand, telephone channels employ a variety of physical media, including wired transmission line, optical fiber cables and microwave. The characteristic of the channel is randomly corrupting the transmitted signal in a variety of possible manner, such as additive *thermal noise* generated by electronic deices, human made noise like automobile ignition noise, atmospheric noise like lightning discharges due to thunderstorm.

The digital demodulator processes the corrupted digital waveform and generates output bit stream. The bit stream is passed then through the channel decoder to reconstruct correct information after removing the redundant bits from the knowledge of encoding scheme used in the channel encoder end.

To get natural (analog) signal at the output, the last and final stage employed is source decoder. Like channel decoding, source decoder also decodes the input signal according to the source encoding scheme applied at the transmitter end. The schemes may be anything out of PCM, Delta Modulation, Adaptive Delta Modulation (ADM), sigma delta modulation (SDM). The performance of the channel decoder-source decoder pair depends upon the statistical prediction of the amount of error introduced per unit time, i.e., bit error rate. The parameter bit error rate (BER) is highly dependent upon the waveform used at the time of source encoding, the modulation scheme applied, etc. When the two extreme end signals (input analog signal at the transmitter and recovered analog signal at the receiver) are compared, the difference is the measure of the total distortion introduced by the digital communication system.

1.6 MATLAB Programs

1.6.1 Time and Frequency Domain Representation of Signals

```
% In CD: ch1_1.m
% Time & frequency domain representation of signals
% Output: Fig. 1.3
% Programed by Apurba Das (Aug,'09)
clc;
clear all;
close all;
t=0:.001:5;
T=2;
f=1/T;
```

```
phi=pi/4;
a=2*cos(2*pi*f*t+phi);
subplot(2,1,1);
plot(t,a);
xlabel('time-->');
ylabel('amplitude-->');
title('Waveform');

f_dom=fft(a);
f_dom=fftshift(f_dom);
% figure;
subplot(2,1,2);
plot(abs(f_dom));
xlabel('frequency-->');
ylabel('amplitude-->');
title('Amplitude Spectrum');

figure;
f=20;
a=a.*sin(2*pi*f*t);
subplot(2,1,1);
plot(t,a);
xlabel('time-->');
ylabel('amplitude-->');
title('Waveform');

f_dom=fft(a);
f_dom=fftshift(f_dom);
% figure;
subplot(2,1,2);
plot(abs(f_dom));
xlabel('frequency-->');
ylabel('amplitude-->');
title('Amplitude Spectrum');
```

1.6.2 CTSV, DTCV, DTDV Signals

```
% In CD: ch1_2.m
% CTCV, DTCV, DTDV signals
% Output: Fig. 1.5
% Programed by Apurba Das (Aug,'09)
clc;
```

```
clear all;
close all;
t=0:.001:5;
T=5;
f=1/T;
a=1.5*(5*sin(2*pi*f*t)+sin(7*pi*f*t));%+4*cos(4*pi*t);
subplot(3,1,1);
plot(t,a,'k');
xlabel('time-->');
ylabel('amplitude-->');
title('CTCV');
subplot(3,1,2);
t1=0:.5:5;
a1=1.5*(5*sin(2*pi*f*t1)+sin(7*pi*f*t1));
stem(t1,a1,'k');
xlabel('time(nT_S)-->');
ylabel('amplitude-->');
title('DTCV');
subplot(3,1,3);
res=20/8; % resolution
[r c]=size(a1);
s=zeros(1,9);
s(1)=-10;
for i=2:9
    s(i)=s(i-1)+2.5;
end;
for i=1:c
    for j=2:9
        if (a1(i)<=s(j) & a1(i)>=s(j-1))
            a2(i)=(s(j-1)+s(j))/2;
        end;
    end;
end;
stem(t1,a1,'k');hold on;
stairs(t1,a2,'k--');
xlabel('time(nT_S)-->');
ylabel('amplitude-->');
title('DTDV');
legend('Sampled signal','Quantized signal');
```

References

1. Das, A., "Line Communication System", New Age International, 2006
2. Proakis, J.,"Digital Communication", McGraw-Hill, 2000

Chapter 2
Waveform Encoding

2.1 Introduction

'Any natural signal is in analog form'. To respect the said statement and to meet the basic requirement of any type of digital signal processing and digital communication, the essential and prior step is converting the electrical form (through transducer) of the natural analog signal into digital form, as digital modulator or any type of digital signal processor does not accept analog signal as its input. Therefore, to consider the digital transmission of analog signal, it is very important to encode the waveform 1st. This process of *waveform encoding* is done through sampling, quantization and encoding; finally the analog information is converted to digital data.

The digitally coded analog signal produces a rugged signal with high immunity to distortion, interference and noise. This source coding also allows the uses of *regenerative repeater* for long distance communication. In the process of quantization, the approximation results in quantization noise and with a target of removing the noise, the bandwidth becomes comparable to the analog signal. Hence, a trade-off between the noise and bandwidth is to be established.

According to the specification in terms of accuracy-bandwidth trade off different schemes of source encoding are employed. The basic procedure of digital transmission of analog signal is coded pulse modulation or PCM. Quantization can be made non-uniformly [4] to catch up intimate details. To reduce the bandwidth in an efficient way, the process of differential pulse code modulation (DPCM) and delta-modulation (DM) [7] are used. In the prediction process of DM, reduction of hardware can be ensured by a nice engineering called as delta-sigma modulation (DSM). At the end of the chapter, the linear prediction is discussed with the help of Genetic Algorithm (GA).

2.2 Pulse Code Modulation (PCM)

Pulse code modulation (PCM) is a digital transmission system of analog signal having an analog to digital converter (ADC) at the input and a digital to analog converter (DAC) at the output.

A. Das, *Digital Communication*, Signals and Communication Technology,
DOI 10.1007/978-3-642-12743-4_2, © Springer-Verlag Berlin Heidelberg 2010

From the previous discussion it is granted that, the analog signals found in nature are essentially continuous time continuous valued signal. To make the signal compatible to feed into digital communication system, the signal needs to be converted to another form which is essentially represented using digits only. This analog-to-digital conversion follows principally three chronological steps namely, (a) Sampling, (b) Quantization and (c) Encoding. By the process of sampling, the time continuity is broken taking care of sufficiently large (almost 1) probability of regeneration. Quantization reshapes each sample height in voltage or in current mode. Here we get defined levels of voltage/current with respect to defined instants of time. That's why this signal obtained after quantization is called as discrete time discrete valued signal. Now, the remaining task is the representation of the defined set of finite number of probable time and voltage/current values. That can be done by mapping each of the elements of the set to a digital code. Automatically an encoder can map the entire signal into digit space.

When the digits are received un-erroneously (for the simplicity of understanding, it's assumed that no errors occur into the communication channel; if error comes, the problem can be taken care of by error detection and correction in channel coding to be discussed in the Chap. 9) at the destination end, formation of the word is essential. Each word is the tag of each quantization level. Next, a synchronized clock oscillating in sampling frequency can segregate the levels, define the time instants respective of the each level and then a discrete time discrete valued signal is obtained here again merely identical to the input signal of the encoder of the transmitter. The signal is obviously a multi-frequency signal where high frequency component arises due to the process of sampling. Now a low pass filter can reconstruct the original signal with some quantization error, which can be minimized by just increasing the number of quantization levels, i.e., by increasing the number of bits to represent each level.

Here a question of bandwidth noise trade-off arises. The objective of communication must play an important role here. There may a priority on accuracy, may be a priority of speed, or there may exist a compromise between these two. Depending upon the user or system requirement, a lot of variations of PCM are designed.

2.2.1 Process of Sampling

The word 'sample' means a small portion of physical element/phenomena which singularly or collectively has all the characteristics of the original physical element/phenomena. As discussed in Chap.1, this is the 1st step towards analog to digital signal conversion. Let's consider the signal m(t) of bandwidth B Hz. When the signal is multiplied by a train of impulses, the resultant signal is obviously the sampled version of m(t). The impulses are equi-spaced by an amount of time T_S. After sampling, the signal g(t) becomes discrete in time domain (Fig. 2.1).

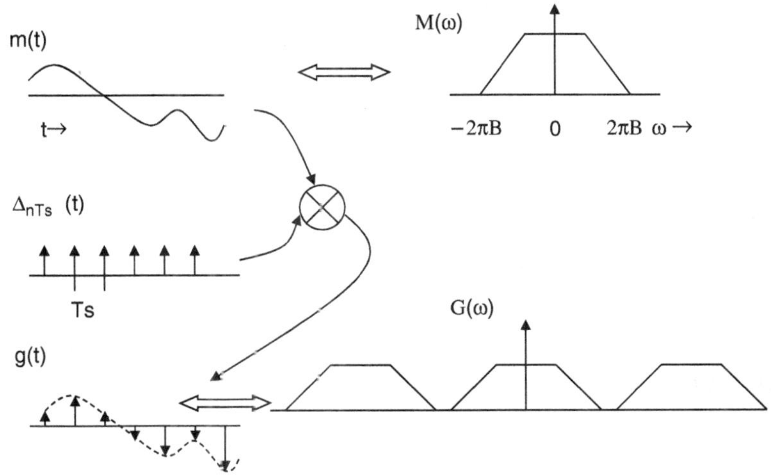

Fig. 2.1 The process of sampling to get discrete time signal

To establish the relationship between the analog signal and discrete (sampled) signal, let's take the example of sampling of a sinusoid s(t), sampled at a rate of F_S, i.e., sampling interval T_S

The analog signal is given by,

$$S(t) = A \sin(\Omega t + \phi) \tag{2.1}$$

where,

A is the amplitude,
Ω is the analog angular frequency $= 2\pi F$,
F is the analog frequency,
ϕ is the initial phase

After sampling, we are taking the signal at $t = 0, T_S, 2T_S, \ldots NT_S$.
Or, t=nT_S, (n=1,2,3,...N), N=number of samples.
Therefore, from Eq. (2.3),

$$\begin{aligned} S(nT_S) &= A \sin(2\pi F n T_S + \phi) \\ \Rightarrow S(n) &= A \sin\left(2\pi \times \frac{F}{F_S} + \phi\right) \text{ as } T_S = 1/F_S \\ \Rightarrow S(n) &= A \sin(2\pi \times f + \phi) \\ \Rightarrow S(n) &= A \sin(\omega n + \phi) \end{aligned} \tag{2.2}$$

taking

$$f = \frac{F}{F_S}$$

From Eq. (4.1) and (4.2), the correspondence of sampling is shown. The sequence formed as S(n) and the input signal S(t) to the sampler, are correspondent.

Looking closely regarding the units of the different variations of frequencies, generated as a bi-product of the process of sampling, we have the following derived units.

Unit of F= cycles/s
Unit of F_S= samples/s

Therefore,

Unit of digital frequency, f=cycles/sample
Unit of analog angular frequency, Ω, i.e., $2\pi F$ = rad.cycles/s
Unit of digital angular frequency ω=rad/samples.

Here, if one decrease the spacing between the samples (T_S), i.e., increase the rate of sampling F_S, the reconstruction would be easier at the receiver side. But, the bandwidth of transmission would be increased thereafter, it will affect the processing time of the sampled signal. So, there is a trade-off between noise and transmission bandwidth. However, it's really important to know the lower limit of the choice of the sampling frequency for successful reconstruction.

2.2.1.1 Sampling Theorem

An analog signal can be reconstructed from its sampled values un-erroneously, if the sampling frequency is at least twice the bandwidth of the analog signal.

Say an analog signal m(t) has three different frequency components f_1, f_2, and f_3, or combination of all of them, where $f_1 < f_2 < f_3$ then the bandwidth of that m(t) signal must be f_3. For ease of calculation, we are taking bandwidth of that signal B Hz, in general.

Therefore we can say the spectrum $M(\omega)$ is band limited by $2\pi B$ in ω scale, and band limited by B Hz in f scale. Now, sampled signal is nothing but the signal obtained by multiplying m(t) by unit impulse train $\Delta_{nTs}(t)$. From the figure the sampled signal is $g(t) = m(t) \times \Delta_{nTs}(t)$. Time period of the impulse train is T_s. therefore frequency is $f_s = 1/T_s$. Let's now expand the $\Delta_{nTs}(t)$ signal in Fourier series so that we can study the spectrum of g(t) signal. As $\Delta_{nT_S}(t)$ is even function of time, By Fourier series,

$$\Delta_{nT_S}(t) = a_0 + \sum_{n=1}^{\infty} (a_n \cos n\omega_s t + b_n \sin n\omega_s t) \tag{2.3}$$

Here,

$$a_0 = \frac{1}{T_S}\int_{T_S} \Delta_{nT_S}(t)\,dt = \frac{1}{T_S}$$
$$a_n = \frac{2}{T_S}\int_{T_S} \Delta_{nT_S}(t)\cos n\omega_0 t dt = \frac{2}{T_S}$$
$$b_n = 0, \text{ for even function}$$

Here, $\int_{T_S}$ signifies integration within upper limit to lower limit difference T_S. Therefore,

$$\begin{aligned}
\Delta_{nT_S}(t) &= \frac{1}{T_S}(1 + 2\cos\omega_s t + 2\cos 2\omega_s t + 2\cos 3\omega_s t + \ldots\ldots\ldots\ldots)\\
\Rightarrow g(t) &= m(t) \times \Delta_{nT_S}(t)\\
&= \frac{1}{T_S}(m(t) + 2m(t)\cos\omega_s t + 2m(t)\cos 2\omega_s t + 2m(t)\cos 3\omega_s t + \ldots\ldots\ldots)\\
&= \frac{1}{T_S}\left(m(t) + 2m(t) \times \frac{e^{j\omega_s t} + e^{-j\omega_s t}}{2} + 2m(t) \times \frac{e^{2j\omega_s t} + e^{-2j\omega_s t}}{2}\right.\\
&\quad \left. + 2m(t) \times \frac{e^{3j\omega_s t} + e^{-3j\omega_s t}}{2} + \ldots\ldots\right)\\
&= \frac{1}{T_S}\left(m(t) + m(t)\left(e^{j\omega_s t} + e^{-j\omega_s t}\right) + m(t)\left(e^{2j\omega_s t} + e^{-2j\omega_s t}\right)\right.\\
&\quad \left. + m(t)\left(e^{3j\omega_s t} + e^{-3j\omega_s t}\right) + \ldots\ldots\right)
\end{aligned} \tag{2.4}$$

From the 1st term of the equation above, it is very clear that spectrum of g(t) will be similar to M(ω) except the amplitude. Amplitude will be 1/ T_s times of that of G(ω). 2nd term indicates M(ω) shifted in both sides by the amount of ω_s (using negative frequency concept). Similarly 3rd term indicates M(ω) shifted in both sides by the amount of $2\omega_s$ and so on.

Now if we can extract the central spectrum by low pass filter from the assembly of spectrum, then we can easily reconstruct the m(t) signal.

Case I

From the Fig. 2.2 above specific fs i.e. ω_s is taken into account. It is very clear that central spectrum can be extracted using LPF. Therefore it is *success* of reconstruction.

Here $\omega_S - 2\pi B = 2\pi B$
Or, $\omega_S = 4\pi B$
Or, $f_s = 2B$

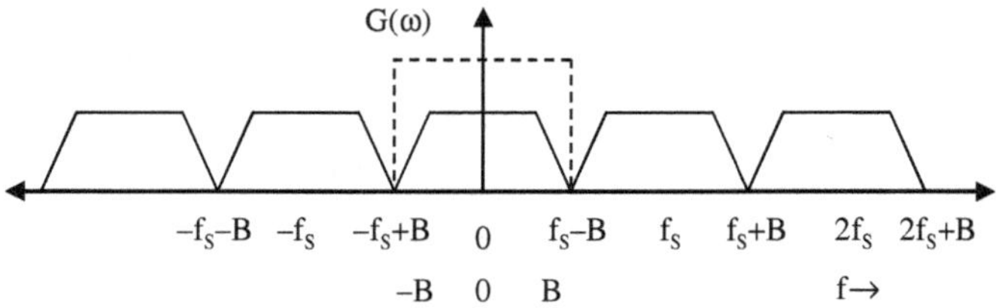

Fig. 2.2 Sampled spectra for f_S=B

Case II

From the Fig. 2.3 above other f_s i.e. ω_S is taken into account. It is very clear that central spectrum cannot be extracted using LPF; because, there is an overlap region of two spectra. The error due to this overlap region is called as *aliasing error*. Therefore it is *failure*of reconstruction.

Here $\omega_S - 2\pi B < 2\pi B$
Or, $\omega_S < 4\pi B$
Or, $f_s < 2B$

Fig. 2.3 Sampled spectra for f_S <B

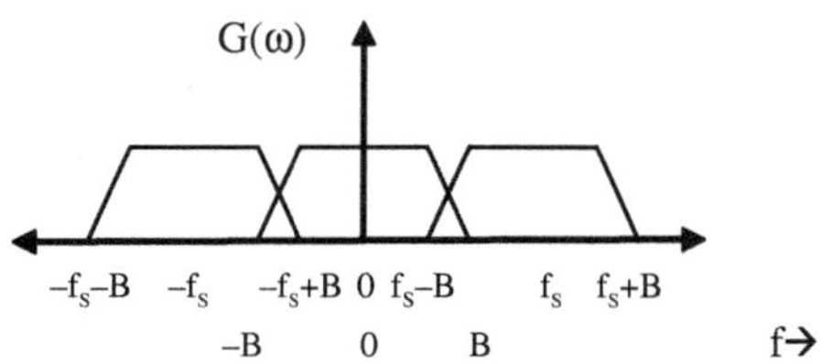

Case III

From the Fig. 2.4 above specific fs i.e. ω_S is taken into account. It is very clear that central spectrum can be extracted using LPF more easily than case I. Therefore it is *success* of reconstruction.

Here $\omega_S - 2\pi B > 2\pi B$
Or, $\omega_S > 4\pi B$
Or, $f_s > 2B$

Therefore collectively we can say $f_s > 2B$ i.e., sampling theorem is proved. The minimum frequency of sampling for successful reconstruction is Nyquist sampling rate ($f_S = 2B$)

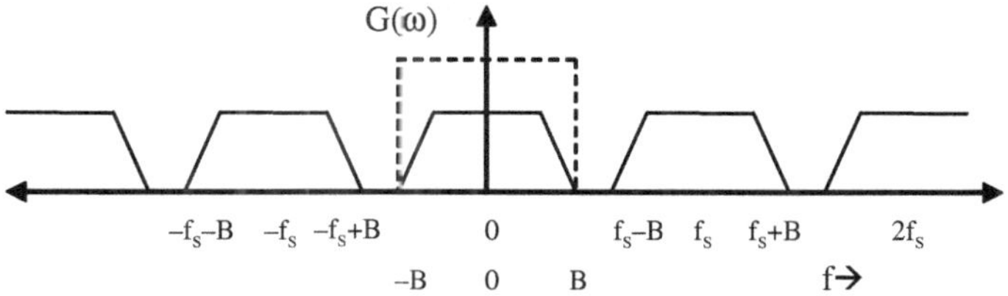

Fig. 2.4 Sampled spectra for $f_S > B$

Collectively, the condition for successful reconstruction of the message signal can be derived therefore as,

$$fs \geq 2B \tag{2.5}$$

2.2.1.2 Aliasing

Say, two sequences are there s1(n) and s2(n). If the digital angular frequency difference between these two sequences is integral multiple of 360°, the sequences would be identical.

Say,

$$\begin{aligned} s1(n) &= A \sin(\omega_1 n + \phi) \\ s2(n) &= A \sin(\omega_2 n + \phi) \end{aligned} \tag{2.6}$$

and

$$\omega_1 = \omega_2 + 2\pi$$

Then,

$$s1(n) = A \sin(\omega_2 n + \phi + 2\pi n) = A \sin(\omega_2 n + \phi) \tag{2.7}$$

It is a severe error. The above equation signifies, after sampling the uniqueness of the signal destroys. Two different sinusoids map to a single sequence after sampling (Fig. 2.5). This error is called as aliasing error and these two sequences are called as alias of other.

To overcome the problem of aliasing, the choice of ω should be such that, any ω must reside within $\pm\pi$. Then only the difference between any two ω's be restricted within the range of 2π as shown in the Fig. 2.6.

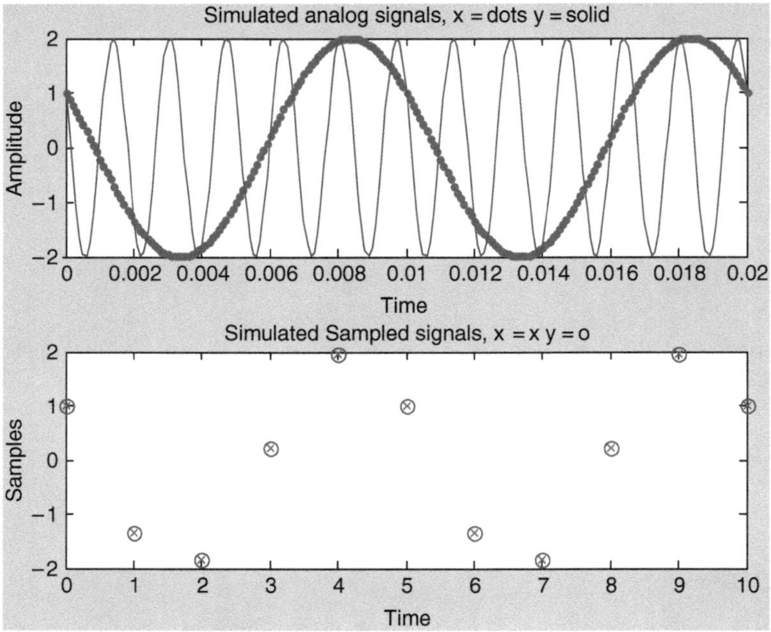

Fig. 2.5 Illustration of the sampling theorem

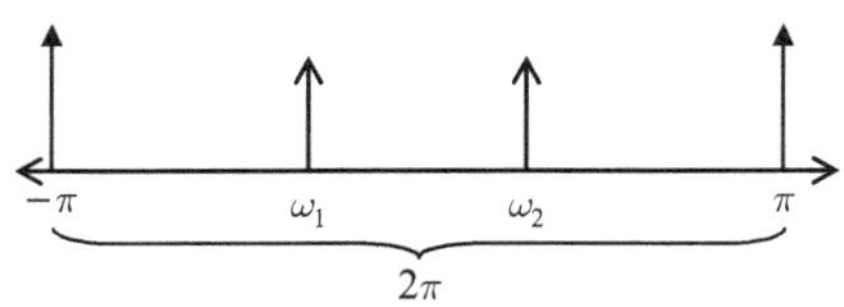

Fig. 2.6 Solution to the error of aliasing

Therefore, for 0 aliasing, the digital angular frequency should be as follows

$$\begin{aligned}
&|\omega| \leq \pi \\
&\Rightarrow 2\pi \frac{F}{F_s} \leq \pi \\
&\Rightarrow F \leq \frac{F_s}{2} \\
&\Rightarrow F_s \leq 2F
\end{aligned} \tag{2.8}$$

Equation (2.8) is indeed the mathematical model for statement of the sampling theorem, and the process of sampled sinusoid signal correspondence is another way to prove the sampling theorem.

2.2.2 Process of Quantization

After sampling what we get is a sequence of impulses with value continuity. The impulses are separated by an amount of the sampling time T_S, but one cannot infer about the amplitude of the signal even after sampling in terms of discontinuity. It means, a set can be defined with finite number of time elements but a set with finite number of probable amplitude elements cannot be defined yet. The amplitude may be anything between the A_{max} and A_{min}. As for an example, if the signal is within voltage range +5 V and −5 V, at an instant of sampling the amplitude may be 3.0000000029 V. And we do not have the control over the accuracy permission, i.e., how many decimal digits we should allow, or truncate or round off and so on. This signal is therefore a discrete time continuous valued (DTCV) signal after sampling. Now, we need to make it properly a discrete time discrete valued (DTDV) signal. The process by which a set of finite number of amplitude values are also defined is called as *quantization.* Say, a set is defined with voltage values 1, 2, 3, 4, 5 and their negative counterpart volts. Then any original value obtained by sampling will be approximated to the closest defined value. For some cases, the sampled values may be quantized high; sometimes they may be quantized low with respect to the defined levels (Fig. 2.7).

To understand closely, let's consider the signal of Fig. 2.8. The maximum and minimum values are designated as A_{max} and A_{min}. Now, the entire range A_{max}

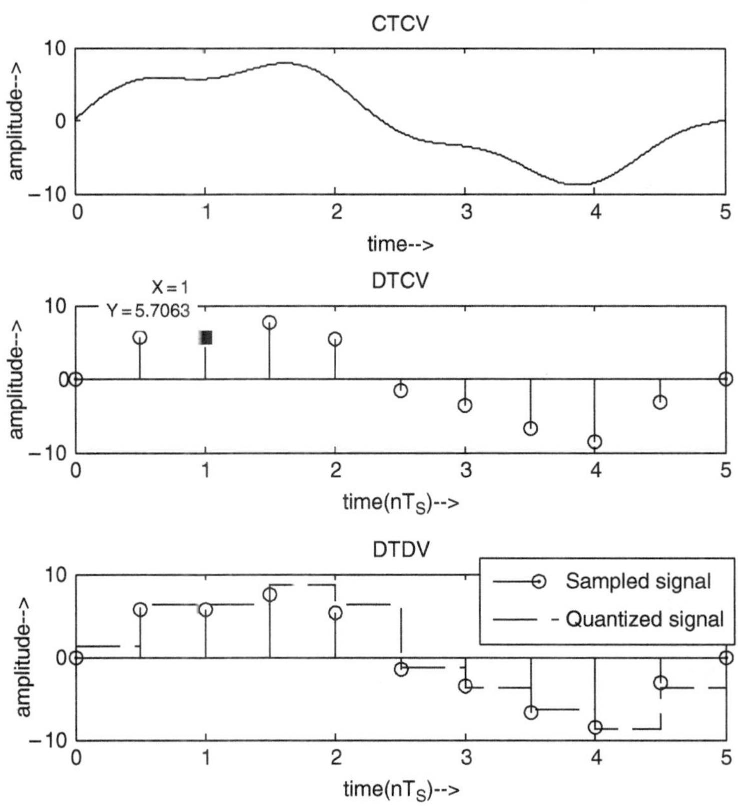

Fig. 2.7 Time valued definition of signal

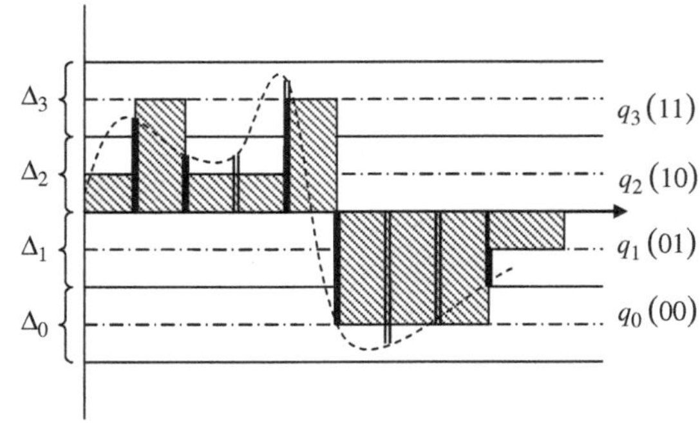

Fig. 2.8 Illustration of uniform quantization and encoding

to A_{min} is equally divided into 4 regions of width S. The 4 regions are named as Δ_0 to Δ_3. The middle of each of the Δ-regions is our defined probable amplitude level. The levels are called as q-levels or quantization levels. The algorithm of quantization states, 'If the sampled signal amplitude lies within Δ_i region, the amplitude will be quantized (approximated) to the amplitude q_i'. Figure 2.8 clearly depicts the process.

The shaded signal in the above figure is the staircase representation of the quantized signal. Next, each q-level can easily be encoded to through 2 bit encoder or MUX to get proper digital representation of the input analog signal. After quantization, the signal $x_q\,(kT_S)$ is represented as $q_2, q_3, q_2, q_2, q_3,\ q_0, q_0, q_0, q_0, q_1$. And after encoding, each level is inferred by a pair of digits in parallel. Next, by employing a parallel to serial converter, the serial bit stream is obtained which is the corresponding digital representation of the input analog signal.

There are two types of uniform quantizer namely,

(a) Midtread quantizer
(b) Midrise quantizer

Figure 2.9 shows the input-output characteristics of the two types of the uniform quantizer. In midtread type quantizer, the origin lies in the middle of the tread of the staircase like graph. In midrise type, the origin the graph lies in the middle of a rising part of the staircase like graph. It can be noticed that, both midtread and midrise graphs are symmetric about the origin.

2.2.3 PCM Transmitter and Receiver

2.2.3.1 PCM Transmitter

The entire principle of analog to digital conversion is now unveiled to us. Therefore, we can just club up all the procedural blocks to construct the PCM transmitter as shown in Fig. 2.10. We have already discussed the functioning of the blocks. First the input analog signal *x*(*t*) is being passed through a low pass filter of bandwidth

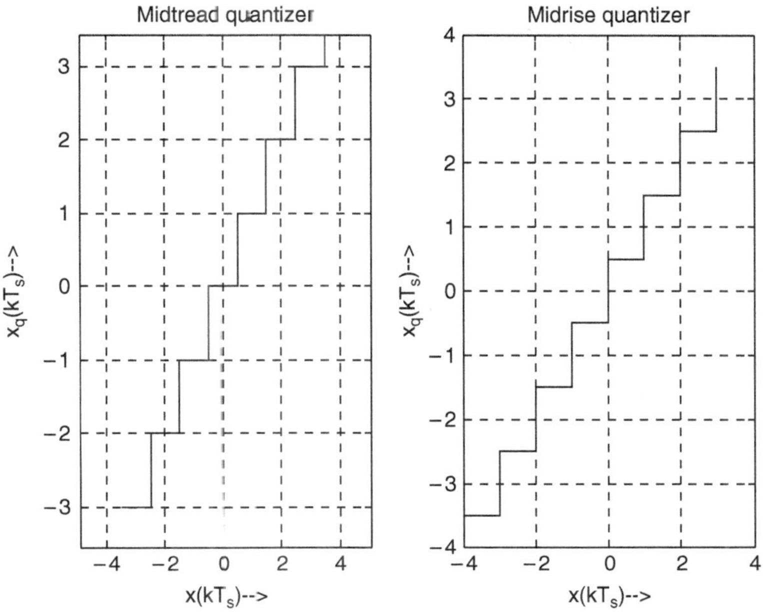

Fig. 2.9 Midtread and midrise quantizer characteristics

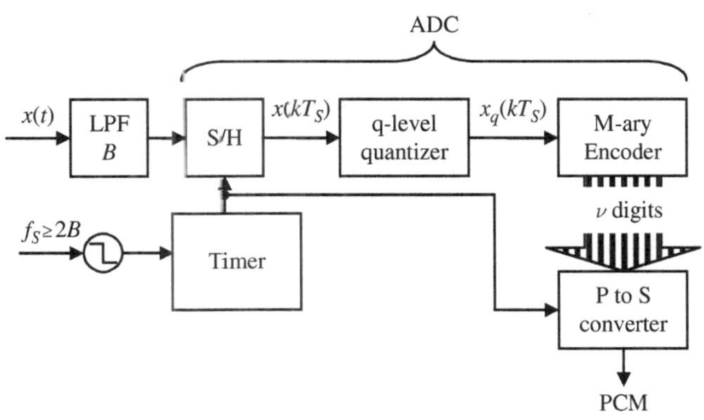

Fig. 2.10 PCM transmitter

B Hz to restrict the bandwidth of the signal for prevention of aliasing error. Next, the three basic steps of analog to digital conversion (ADC) is being followed i.e., sampling-quantization-encoding, as discussed before. But here the encoded bit stream is essentially parallel in form, although the objective is to get serial bit stream in response to serial analog signal input [1].

The encoder works with M-ary digits (if M=2, it becomes binary) and produces a codeword consisting of ν digits in parallel, corresponding to each sample. Since there are M^{ν} (in binary, 2^{ν}) possible M-ary codewords, with ν digits per word, unique coding of the q different quantization levels requires that, $M^{\nu} \geq q$. The parameters should be chosen removing the inequality, such that,

$$q = M^{\nu} \text{ or,} \quad \nu = \log_M q \tag{2.9}$$

For binary PCM, number of bits per q-level can be calculated as $\nu = \log_2 q$.

Finally, the successive code words are read out serially. The parallel to serial conversion also needs the clocking by sampling frequency. Otherwise, either one word would superimpose or overwrite the other, or one code word would be converted twice. In both the cases, we get improper conversion. Therefore, clocking in the parallel to serial converter is essential.

As, each encoded sample is represented using ν digits, so the signaling rate at the output becomes $r = \nu f_S$. Therefore, for transmission of PCM signal, the bandwidth needed is

$$B.W._{tr} \geq \frac{1}{2} r = \frac{1}{2} \nu f_S \geq \nu B \tag{2.10}$$

2.2.3.2 PCM Receiver

The transmitted PCM signal is now passed through transmission channel towards the receiver. Contamination of noise is almost obvious during the traversal of the signal through transmission path. Though the received signal is noise accumulated, the process of regeneration yields a nearly errorless signal as the SNR becomes sufficiently large. In the receiver section ultimately we need to get the reconstructed analog signal. Therefore, one digital to analog conversion (DAC) is to be designed. As shown in the Fig. 2.11, the DAC operation is just the opposite, i.e., serial to parallel conversion (for getting proper form of input to the next stage, i.e. decoder), M-ary decoding and sample and hold. The output of the sample and hold circuit is the staircase type waveform, $x_q(t)$.

The waveform shown in Fig. 2.12 may be regarded as a staircase approximation $x(t)$. A low pass filter is then employed for smoothening. The low pass filtering produces the reconstructed analog signal $\tilde{x}(t)$, which differs from $x(t)$ to the extent that the quantized samples $\left(x_q\,(kT_S)\right)$ differ from the exact sampled values $(x\,(kT_S))$.

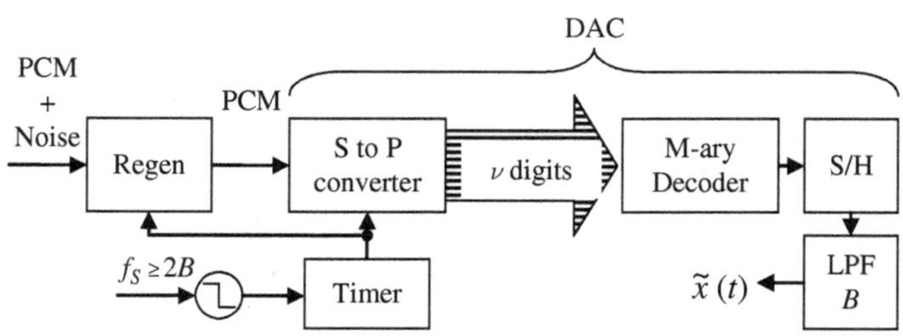

Fig. 2.11 PCM receiver

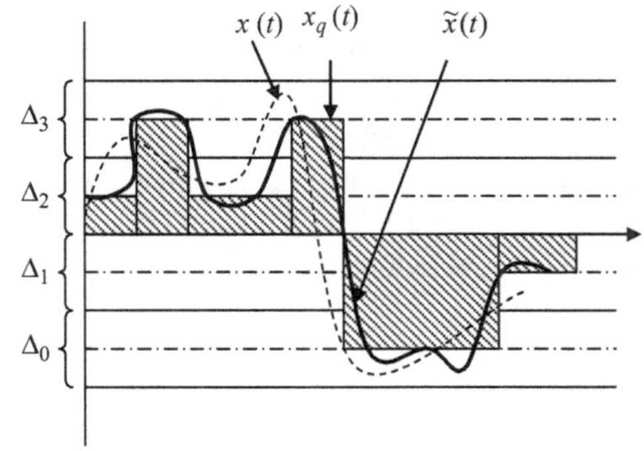

Fig. 2.12 Waveform reconstruction through PCM receiver as shown in Fig. 2.11

2.2.4 Quantization Error

It is now understood that, the process of quantization is nothing but approximation to nearer level of voltage/current. Due to this approximation, there a random amount of difference occurs between actual and quantized value, named as quantization error. As the error is like a random number and adds noise to modulation process, the error can also be viewed as a noise called as *quantization noise* Fig. 2.13.

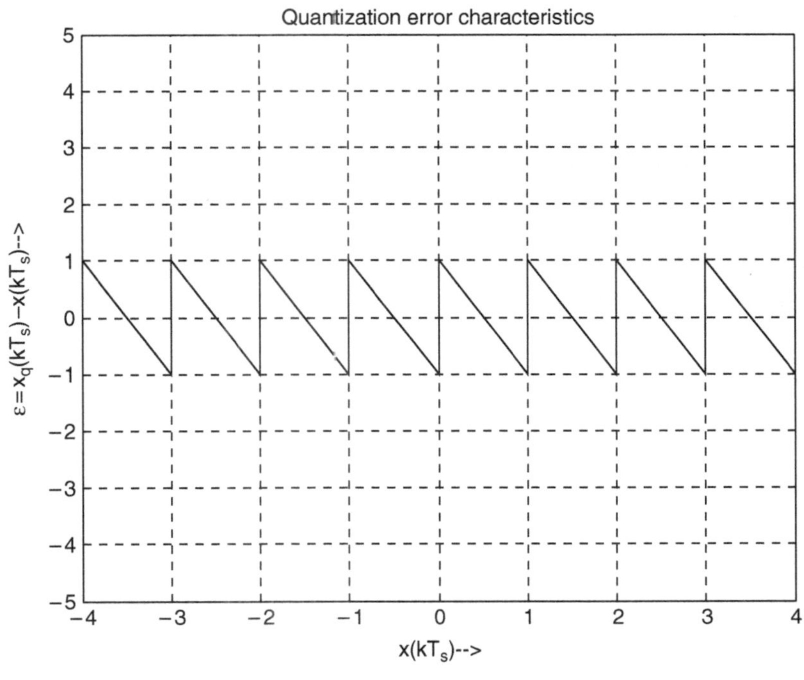

Fig. 2.13 Quantization error characteristics

The noise is now to be calculated as mean-square quantization error, e^2 where e is the difference between the actual and approximated value of voltage/current.

The peak to peak amplitude of the sampled signal $x_s(t)$ is divided into M equal levels each of width S. At the centre of each level of width the quantization levels are located as $x_1, x_2, \dots x_M$ as shown the Fig. 2.14. Since, in the figure $x_s(t)$ happens to be the closest to the level x_k, the quantizer output will be x_k and obviously, the quantization error is $e = x_s(t) - x_k$.

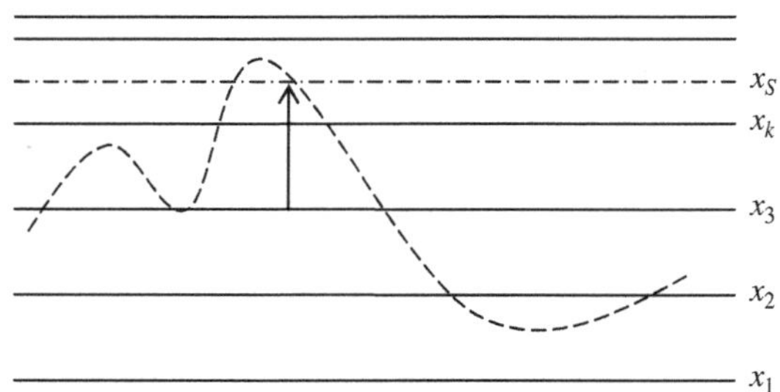

Fig. 2.14 Interpretation of quantization error

Let, $p(x)\,dx$ be the probability that $x_s(t)$ lies in the voltage/current range $x+dx/2$ to $x-dx/2$. Then the mean square quantization error [3] is

$$\overline{e^2} = \int_{x_1-S/2}^{x_1+S/2} p(x)\,(x-x_1)^2\,dx + \int_{x_2-S/2}^{x_2+S/2} p(x)\,(x-x_2)^2\,dx + \dots \tag{2.11}$$

Now, the PDF (probability density function) $p(x)$ of the message signal will not certainly be constant throughout each division. Assuming large M, i.e., with sufficiently small S, $p(x)$ can be taken as constant throughout each division. Then, the 1st term of the right hand side of the Eq. (2.11), $p(x) = p^{(1)}$=constant. The 2nd term is $p(x) = p^{(2)}$, and so on. Hence, the constant terms may be taken out of the integration sign. If we now substitute $y \equiv x - x_k$, the expression in Eq. (2.11) becomes

$$\begin{aligned}\overline{e^2} &= \left(p^{(1)} + p^{(2)} + \dots\right) \int_{-S/2}^{S/2} y^2 dy \\ &= \left(p^{(1)} + p^{(2)} + \dots\right) \frac{S^3}{12} \\ &= \left(p^{(1)}S + p^{(2)}S + \dots\right) \frac{S^2}{12}\end{aligned} \tag{2.12}$$

Now, according to definition, $p^{(1)}S$ be the probability that the signal lies within the 1st quantization range, $p^{(2)}S$ be the probability that the signal lies within the 2nd quantization range, and so on. Hence, the sum of the terms in the bracket in the

Eq. (2.12) must have unity value, as signal must lie within the peak-to-peak value, i.e., summation of all quantization range. Therefore, the mean square quantization error is

$$\overline{e^2} = \frac{S^2}{12} \tag{2.13}$$

2.2.5 Signal to Noise Ratio (SNR) for Quantized Pulses

A signal with maximum and minimum amplitudes as +A and –A is assumed. The signal is quantized in M levels with each of width S. Then the peak to peak range of amplitude (voltage or current) is A_{pp}=*A*– (–*A*) =2*A*. As understood from the graph shown in Fig. 2.13, the maximum value of quantization error is S/2 or –S/2 in either direction. Therefore, the degradation of the signal is limited to the amount of S/2 in additive or subtractive manner. It is obvious that, the peak to peak amplitude can also be expressed in terms of number and width or quantization levels as $A_{pp}=M\times S$.

The peak power of the analog signal (normalized to 1 Ω) can be expressed as

$$A_p{}^2 = \left(\frac{A_{pp}}{2}\right)^2 = \left(\frac{MS}{2}\right)^2 = \frac{M^2S^2}{4} \tag{2.14}$$

Now, we do have already the expression for quantization noise. Therefore, from Eq. 2.13 and Eq. 2.14 we can derive the signal to noise ratio for uniformly quantized pulses as [6]

$$\frac{S_t}{N_q} = \frac{\frac{M^2S^2}{4}}{\frac{S^2}{12}} = 3\,M^2 \tag{2.15}$$

The above equation supports our common understanding that, SNR will be improved as the number of levels to be squared is increased. As a limit if $M \to \infty$, the PCM signals will be converted to PAM signal, i.e., no quantization will be done.

Example 2.1 Consider an audio signal as given
$s(t) = 3\cos 500\pi t$ V

(i) Find the SNR when *s(t)* is quantized using 10 bit PCM.
(ii) How many bits of quantization is needed for achieving SNR of 40 dB.

(i) $SNR = \frac{S_t}{N} = \frac{12P_S}{S^2}$,

where P_S is the RMS signal power defined as,

$$
\begin{aligned}
P_S &= \frac{3^2}{2} = 4.5\,\mathrm{W} \\
S &= \frac{V_H - V_L}{M} = \frac{3-(-3)}{2\times 10^{10}} = 5.86\,\mathrm{mV} \\
\therefore SNR &= \frac{S_t}{N} = \frac{12\times 4.5}{(5.46\times 10^{-3})^2} = 1.56\times 10^6 \\
\Rightarrow SNR\,|_{\mathrm{dB}} &= \left.\frac{S_t}{N}\right|_{\mathrm{dB}} = 10\log\left(1.57\times 10^6\right) \approx 62\,\mathrm{dB}
\end{aligned}
$$

(ii) We know, $40\,\mathrm{dB} = 4\times 10\log 10 = 10\log 10^4$
Therefore, to find out the number of bits of quantization,

$$
\begin{aligned}
\frac{12\times 4.5}{(6/2^\nu)^2} &= 10^4 \\
\Rightarrow \frac{2^{2\nu}}{(6)^2} &= \frac{10^4}{12\times 4.5} \\
\Rightarrow 2^{2\nu} &= \frac{36\times 10^4}{12\times 4.5}
\end{aligned}
$$

Taking $\log_2$ of both sides, we get

$$
\begin{aligned}
\Rightarrow 2\nu &= \log_2\left(\frac{36\times 10^4}{12\times 4.5}\right) = 12.7027 \\
\therefore \nu &= \left\lceil \frac{12.7}{2} \right\rceil = 7\,\mathrm{bits}
\end{aligned}
$$

2.2.6 Non-uniform Quantization: Companding

For most of the voice communication channels, very low speech volumes predominate for more than half of the times, the voltage characterizing detected speech energy is 25% of the RMS value. Large amplitude values are relatively rare; only 10–15% of the time the voltage exceeds the RMS value. This is a peculiar statistics of speech signals. Therefore, the spacing at the high amplitude level becomes an wastage and for incorporation of that unused q-levels, we do have to go for high bandwidth transmission of low accuracy transmission, if we employ uniform quantization.

In case of the design of uniform quantization, the quantization error varies directly proportional to the square of the uniform width of quantization levels, i.e., the quantization error is inversely proportional to the square of number of quantization levels. Now, to reduce the error, if we plan to increase the number of q-levels, it requires more number of bits to represent each level. Hence, the transmission bandwidth will be increased significantly. To bypass the trade-off between these two (transmission bandwidth and quantization error), an overall reduction of quantization error is achieved by judicially choosing the widths (non-uniformly) of the levels keeping the total number of quantization levels unchanged [4, 5].

The ratio of voltage levels covered by voice signals from peaks of loud talk to the weak passages of weak talk is of the order of 1000:1. Such variation is taken care of by *non-uniform quantization* of the signal. The non-uniform quantization helps the practical variability is such a way that, the step size automatically increases as the separation from the origin of the input-output amplitude characteristics is increased. In weak passages, the step size becomes smaller and the details are not missed at all. On the other hand, by employing uniform quantization, either number of steps should be made higher or the weaker signal details have to be compromised.

The procedure for getting non-uniformly quantized samples is compression followed by uniform quantization. At the receiver side also, the reverse operation, i.e., expanding followed by uniform quantization is done, as shown in Fig. 2.15. The inclusion of COMpression and exPANDING renames the process of non-uniform quantization as a contraction COMPANDING (Fig. 2.16).

It does can be easily understood that, the compression characteristic and parameter of the characteristic is really an important factor to know for reliable and successful communication. A typical compressor characteristic is shown in

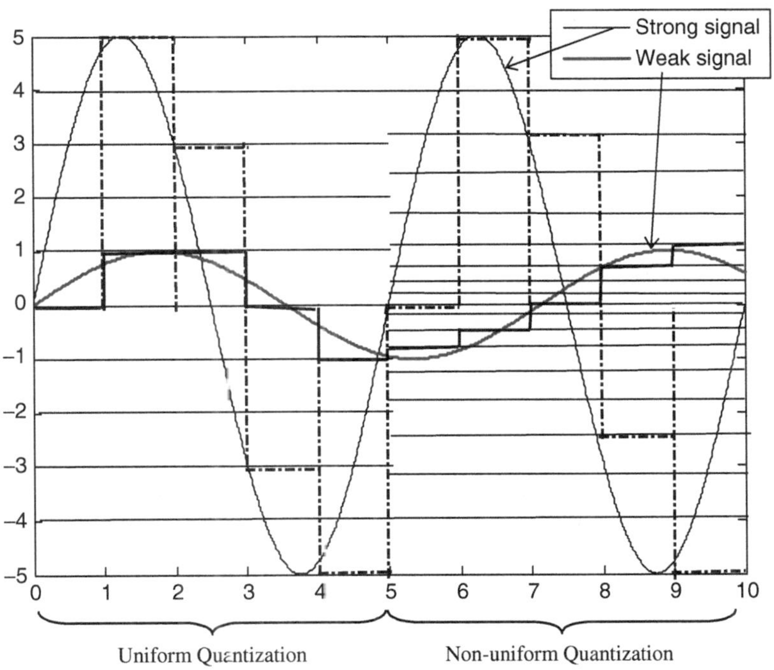

Fig. 2.15 Uniform quantization vs. non-uniform quantization

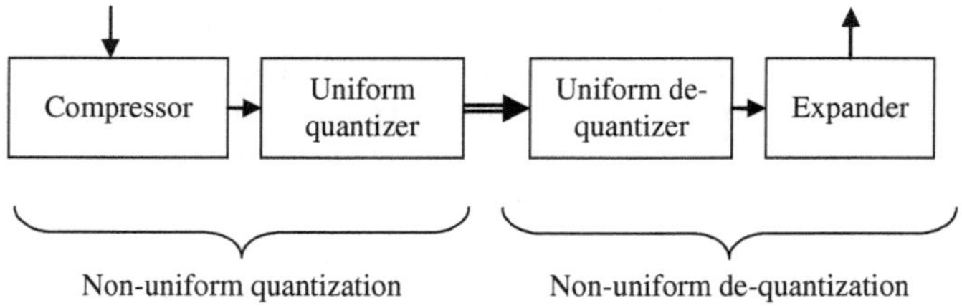

Fig. 2.16 The process of COMPANDING

Fig. 2.17 Compressor characteristics

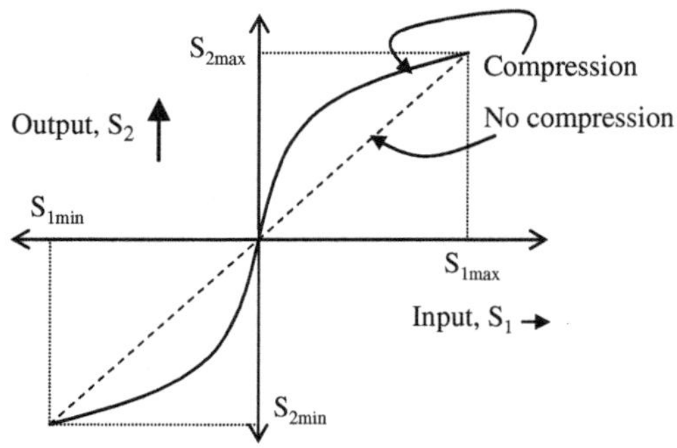

Fig. 2.17. A lot of compressor laws with different parameter orientation are there. Among them A-law and μ law are accepted by CCITT[1] and are discussed here.

μ law are mainly used in North America and Japan. A-law is used in Europe and rest of the country.

2.2.6.1 μ-Law

The compression algorithm following μ-law is given by

$$|s_2| = \frac{\log(1 + \mu|s_1|)}{\log(1 + \mu)} \tag{2.16}$$

where, s_1 and s_2 are normalized input and output voltage/current respectively. μ is a non-negative parameter which determine the degree of compression. For $\mu = 0$, we obtain uniform quantization. The plot of the normalized output with normalized

[1] CCITT is the contraction of Consultative Committee for International Telegraphy and Telephony.

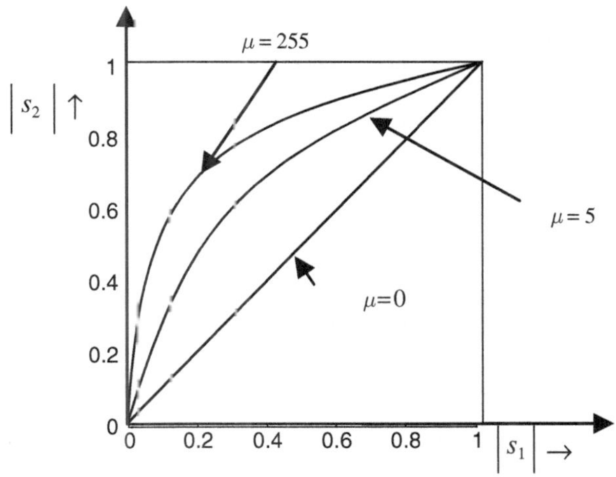

Fig. 2.18 Compression by μ-law

input with different values of μ is shown in Fig. 2.18. The quantization step is defined as the reciprocal of the slope of the compression curve given by

$$\frac{d\,|s_1|}{d\,|s_2|} = \frac{\log(1+\mu)}{\mu}(1+\mu\,|s_1|) \tag{2.17}$$

For lower values of input, $\mu\,|s_1| << 1$, therefore,

$$|s_2| \approx \frac{\mu\,|s_1|}{\log(1+\mu)} \tag{2.18}$$

and

$$\frac{d\,|s_1|}{d\,|s_2|} \approx \frac{\log(1+\mu)}{\mu} \tag{2.19}$$

For higher values of input, $\mu\,|s_1| >> 1$, therefore,

$$|s_2| \approx \frac{\log(\mu\,|s_1|)}{\log(1+\mu)} \tag{2.20}$$

and

$$\frac{d\,|s_1|}{d\,|s_2|} \approx |s_1| \log(1+\mu) \tag{2.21}$$

Therefore, the μ-law is linear at low input levels and approximately logarithmic at high input levels.

2.2.6.2 A-Law

Another well known compression algorithm following A-law is given by

$$|s_2| = \frac{A\,|s_1|}{1+\log A}; \qquad 0 \le |s_1| \le \frac{1}{A} \tag{2.22}$$

$$|s_2| = \frac{1+\log\left(A\,|s_1|\right)}{1+\log A}; \qquad \frac{1}{A} \le |s_1| \le 1 \tag{2.23}$$

Here also, A is a non-negative parameter of compression. A=1 signifies uniform-quantization. The plot of the normalized output with normalized input with different values of A is shown in Fig. 2.19. The quantization step is defined as the reciprocal of the slope of the compression curve given by

$$\frac{d\,|s_1|}{d\,|s_2|} = \begin{cases} \dfrac{1+\log A}{A}; & 0 \le |s_1| \le \dfrac{1}{A} \\ (1+\log A)\,|s_1|\,; & \dfrac{1}{A} \le |s_1| \le 1 \end{cases} \tag{2.24}$$

Thus, the quantization steps over the central linear segment are diminished by factor A/(1+logA). These steps over the linear segment have a dominant effect on small signals and are diminished practically by about 25 dB as compared to uniform quantization.

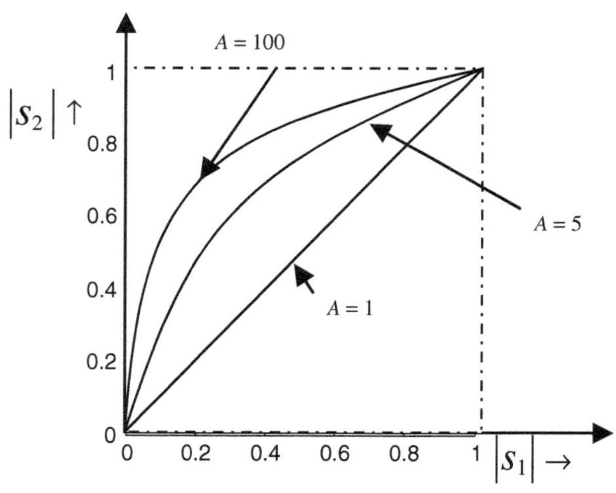

Fig. 2.19 Compression by A-law

Example 2.2 When a 256 level quantization is employed for a sinusoid with peak voltage 40 Volts, what is the voltage interval without compression?

For $\mu = 255$, what is the maximum and minimum separation between the levels?

(i) For uniform quantization, the step size, S can be determined easily as

$$S = \frac{2 \times V}{256} = \frac{2 \times 40}{256} = 312.5\,\text{mV}$$

(ii) To apply μ-*law* for non-uniform quantization, for $\mu = 255$
The smallest height h would be increased in GP as

$$h + 2h + 4h + 8h + 16h + 32h + 64h + 128h = 2\,\text{V}$$
$$\Rightarrow 255\,h = 2\,\text{V}$$
$$\Rightarrow h = 0.0078\,\text{V} = 7.8\,\text{mV}$$
$$\Rightarrow 128\,h = 1.0039\,\text{V}$$

Therefore the maximum and minimum step size employing μ-*law* for non-uniform quantization, for $\mu = 255$ are 1.0039 V and 7.8 mV respectively.

2.3 Differential Pulse Code Modulation (DPCM)

In analog messages we can make a good guess about a sampled value from the knowledge of the past sampled values. In other words, the sampled values are not independent, and generally there is a great deal of redundancy in the Nyquist samples. Proper exploitation of this redundancy leads to encoding a signal with lesser number of bits. Consider a sampling scheme where instead of transmitting the sampled values, we transmit the difference between the successive samples. By employing the technique of transmitting the quantized difference values of the successive samples we can efficiently use the bandwidth provided by the transmitting channel.

2.3.1 Cumulative Error in PCM

In general PCM system, a quantization error e_{q_i} is added to the quantized output while quantizing the ith sample. Now just take a look on how the quantization error affects the DPCM output.
Here Z_i is the sampled value at ith instant and $\hat{Z}_i$ is the predicted sample (considered as delayed sample) at ith instant.

Now, from the Fig. 2.20,

$$\begin{aligned} d_i &= Z_i - \hat{Z}_i \\ &= Z_i - Z_{i-1} \end{aligned} \tag{2.25}$$

Or recovered sample at the receiver is $\overline{Z}_i = \hat{d}_i + \overline{Z}_{i-1}$

For the 1st sample, i=1. So, $\overline{Z}_1 = \hat{d}_1 + \overline{Z}_0$

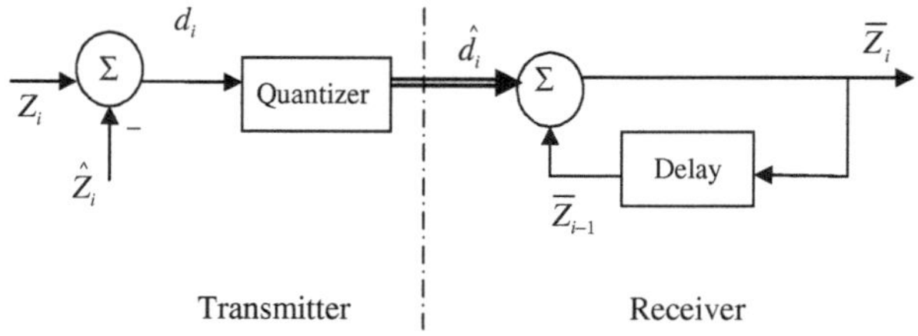

Fig. 2.20 Basic DPCM communication system

$$= \hat{d}_1 + Z_0 (\text{As at 0th instant}\, \overline{Z}_0 = Z_0)$$

And the quantized difference

$$\hat{d}_i = d_i + e_{qi} \tag{2.26}$$

So, from the diagram presented in Fig. 2.20,

$$\begin{aligned} \overline{Z}_1 &= d_1 + e_{q1} + Z_0 \\ \text{Or } \overline{Z}_1 &= Z_1 + e_{q1} \end{aligned} \tag{2.27}$$

Now, for i=2, we get from Eq. (2.25),

$$\begin{aligned} \overline{Z}_2 &= \hat{d}_2 + \overline{Z}_1 \\ \overline{Z}_2 &= d_2 + e_{q2} + Z_1 + e_{q1} \\ \overline{Z}_2 &= Z_2 + e_{q1} + e_{q2} \end{aligned} \tag{2.28}$$

Proceeding in this way we can see

$$\overline{Z}_n = Z_n + \sum_{i=1}^{n} e_{qi} \tag{2.29}$$

So, here we see that in case of basic configuration of DPCM, the output at nth instant is affected not only by the quantization error of nth instant, but also by the errors of previous instants. The error becomes severe as the term of samples increases. To overcome the problem, a new configuration is designed.

2.3.2 Prevention of Cumulative Error by Applying Feedback

In the receiver section, a delay feedback path is used for proper recovery of the message sample. To solve the problem of cumulative quantization error in the transmitted samples, we have just copied the feedback loop of the receiver section and pasted to the transmitter section as shown in Fig. 2.21.

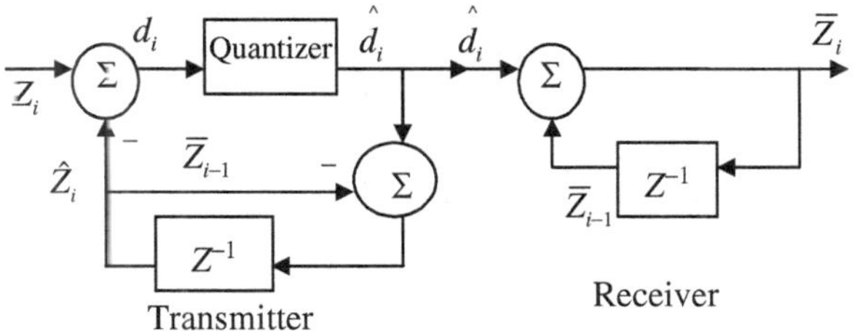

Fig. 2.21 DPCM communication system with error suppressor

feedback circuit has been added to the transmitter side. The feedback circuit is basically same as the receiver circuit. Now let us analyze the transmitter. Here the block Z^{-1}signifies 1 unit delay (delay by an amount of T_S).

At ith instant

$$d_i = Z_i - \overline{Z}_{i-1} \text{ and } \hat{d}_i + \overline{Z}_{i-1} = \overline{Z}_i \tag{2.30}$$

Also,

$$\hat{d}_i = d_i + e_{q_i} \tag{2.31}$$

For the 1st sample, i.e. for i=1,

$$\begin{aligned} \hat{d}_1 &= d_1 + e_{q1} \\ \text{So, } \overline{Z}_1 &= \hat{d}_1 + \overline{Z}_0 \\ &= d_1 + \overline{Z}_0 + e_{q_1} \\ &= Z_1 + e_{q_1} \end{aligned} \tag{2.32}$$

For i=2, from Eq. (2.31) we get

$$\hat{d}_2 = d_2 + e_{q2} \tag{2.33}$$

Now,

$$\hat{d}_2 + \overline{Z}_1 = \overline{Z}_2$$

So,

$$\begin{aligned} \overline{Z}_2 &= d_2 + e_{q_2} + \overline{Z}_1 \\ &= Z_2 + e_{q_2} \end{aligned} \tag{2.34}$$

Proceeding in this way, we can say reconstructed output at any instant

$$\overline{Z}_n = Z_n + e_{q_n} \tag{2.35}$$

So, comparing Eqs. (2.29) and (2.35), we can say that in the present case, the reconstructed output depends only on the quantization error of that particular instant. So it's free from cumulative error.

Thus if $m[k]$ is the kth sample instead of transmitting $m[k]$ we transmit the difference $d[k] = m[k] - m[k-1]$. At the receiver knowing $d[k]$ and the previous sample $m[k-1]$, we reconstruct $m[k]$ iteratively at the receiver. Now, the differences between successive samples are generally much smaller than the sample values. Thus the peak amplitude m_p of the transmitted value is reduced considerably because the quantization interval $A_v = m_p/l$, for a given l (no of bits per sample) we can reduce the quantization noise given by $A_v^2/12$. This also means that for a given transmission bandwidth (which is related to number of bits per sample as $BW = \frac{l \times \text{samples/s}}{2}$) and for a given SNR we can reduce BW.

We can improve upon this scheme by estimating (predicting) the value of kth sample $\hat{m}[k]$ from a knowledge of the previous sample. If this estimate is $\hat{m}[k]$, then we transmit the difference (prediction error) $d[k] = m[k] - \hat{m}[k]$). At the receiver also, we determine the estimated $\hat{m}[k]$ from the previous sample, and then generate $m[k]$ by adding the received $\hat{d}[k]$ to the estimated $\hat{m}[k]$.

Now if our prediction is quite worth, the difference between $\hat{m}[k]$ and $m[k]$ will be much smaller than the difference between two sample values $m[k]$ and $m[k-1]$. Thus the number of bits required will be even lower than the previous case and hence the bandwidth will also be lower.

In the basic assumption, the prediction for a particular sample is taken granted as the delayed sample. In the next section, we'll try to discuss the reason and logic behind this assumption.

2.3.3 How We Can Predict the Future?

The approach to signal prediction (estimation) is discussed here using the spirit of Taylor [2].

Let us consider, a signal $m(t)$, which have derivatives of all order at t. Using Taylor series we can express $m(t + T_S)$ as

$$m(t + T_S) = m(t) + \frac{T_S}{1!}m(t) + \frac{T_S^2}{2!}m(t) + \ldots\ldots\ldots\ldots \tag{2.36a}$$

$$\approx m(t) + T_S m(t) \text{ For small } T_S \tag{2.36b}$$

Equation (2.36a) shows that from knowledge of the signal and its derivative at instant t, we can predict a future signal at $t = T_S$. In fact, even if we know just the first derivative, we can still predict this value approximately, as shown in

Eq. (2.36b). Let us denote the kth sample of $m(t)$ as $m[k]$, that is $m(kT_S) = m[k]$, and $m(kT_S \pm T_S) = m[k \pm 1]$.

Now, the derivative

$$m(kT_S) \approx \frac{m[kT_S] - m(kT_S - T_S)}{T_S}.$$

So form Eq. (2.36b), we obtain

$$\begin{aligned} m[k+1] &\approx m[k] + T_S \left[\frac{m[k] - m[k-1]}{T_S} \right] \\ &= 2m[k] - m[k-1]. \end{aligned} \tag{2.37}$$

It shows that, we can find a crude prediction of the $k + 1$th sample from the two previous samples. The approximation improves if we take more terms in Eq. (2.36a). To determine the higher order derivatives in the services, we require more samples from the past. Larger the number of past samples, we can get a better approximation.

$$m[k] \approx a_1 m[k-1] + a_2 m[k-2] + \cdots\cdots + a_N m[k-N] \tag{2.38}$$

The right hand side is $\hat{m}[k]$ the predicted value of $m[k]$. Thus $\hat{m}[k] = a_1 m[k-1] + a_2 m[k-2] + \cdots + a_N m[k-N]$.

This is the equation of an Nth order prediction. Thus we can design one prediction filter whose output is $\hat{m}[k]$ and input is $m[k-1], m[k-2], \ldots m[k-N]$. But we can get these by simply delaying $m[k]$ by T_S.

The parameters a1, a2, . . ., aN are called prediction coefficients. This is also called linier prediction. It is actually a transversal filter (a tapped delay line), where the tap gains are set equal to the prediction coefficients as shown in Fig. 2.22.

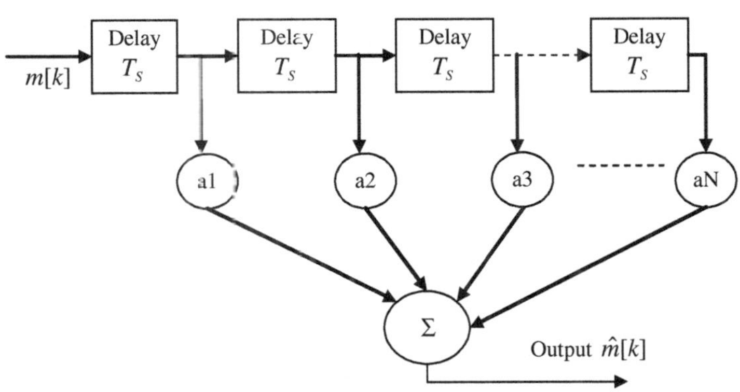

Fig. 2.22 Basic idea of linear prediction

2.3.4 Analysis of DPCM

DPCM transmit $d[k]$, the difference between $m[k]$ and $\hat{m}[k]$ and at receiver we generate $\hat{m}[k]$ from the past sample values to which the received $d[k]$ is added to generate $m[k]$.

There is, however, one difficulty in this scheme. At the receiver, instead of past sample $m[k-1]$, $m[k-2]$, ... $m[k-N]$ as well as $d[k]$, we have there quantized versions, $m_q[k-2]$,... $m_q[k-N]$. Hence, we cannot determine $\hat{m}[k]$. We can only determine $\hat{m}_q[k]$ the predicted value of quantized sample from previous quantized samples. This will increase the error in reconstruction (because the error is here additive, error in each quantized signal is added together).

In such case, a better strategy is to determine $\hat{m}_q[k]$, the estimate of $m_q[k]$ instead of $m[k]$ at the transmitter also from the quantized sample. The difference $d[k] = m[k] - \hat{m}_q[k]$ is now transmitted using PCM (Fig. 2.23). At the receiver, we can generate $\hat{m}_q[k]$, and from the received $d[k]$, we can reconstruct $m_q[k]$ (Fig. 2.24).

Here $d_q[k] = d[k] + q[k]$, where $q[k]$ is the quantization error. The predictor output $\hat{m}_q[k]$ is fed back to its input so that the predictor input $m_q[k]$ is $m_q[k] = \hat{m}_q[k] + d_q[k]$

$$
\begin{aligned}
&= m[k] - d[k] + d_q[k] \\
&= m[k] + q[k]
\end{aligned}
\tag{2.39}
$$

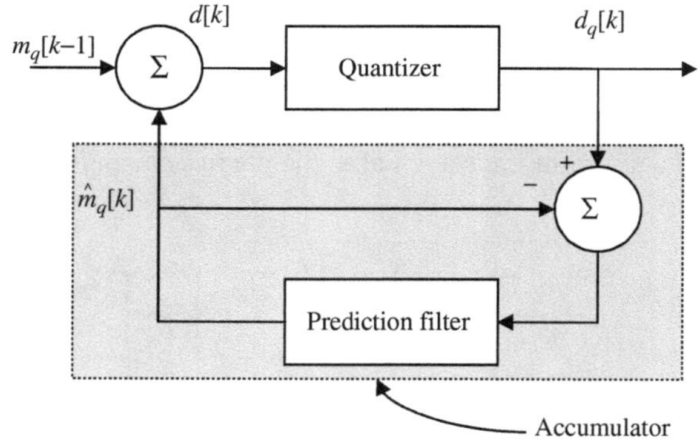

Fig. 2.23 DPCM transmitter

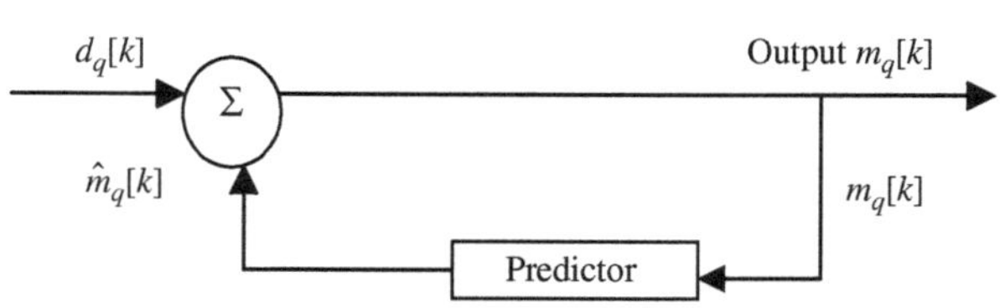

Fig. 2.24 DPCM receiver

This shows that $m_q[k]$ is a quantized form of $m[k]$. The predictor input is indeed $m_q[k]$ as assumed.

The receiver shown in Fig. 2.24 is identical to the shaded portion of the transmitter. The inputs in both cases are also the same $d_q[k]$. The predictor output must be $m_q[k]$.

2.4 Delta Modulation

Sample correlation used in DPCM is further explained in Delta Modulation (DM) by over sampling (typically 4 times the Nyquist rate) the baseband signal. This increases the correlation between adjacent samples, which results in a small prediction error that can be encoded using only one bit (no. of levels = 2). Thus DM is basically a one bit DPCM, that is, a DPCM that uses only two levels (L=2) for quantization of the $m[k] - \hat{m}_q[k]$. For more simple scheme the difference between $m[k]$ and $m[k-1]$ is encoded by only one bit. That is if $m[k]$ is higher than $m[k]$ then 1 is transmitted, otherwise 0.

In comparison to PCM (and also DPCM), it is very simple method of A/D conversion. In DM the word framing is unnecessary in transmitter and receiver. This strategy allows us to use fewer bits per sample for encoding a base band signal.

In delta modulation, we use a first order predictor, which, as seen earlier, is just a time delay of T_S. Thus the DPCM transmitter and receiver can be reconfigured to delta modulation as shown in the following Fig. 2.25.

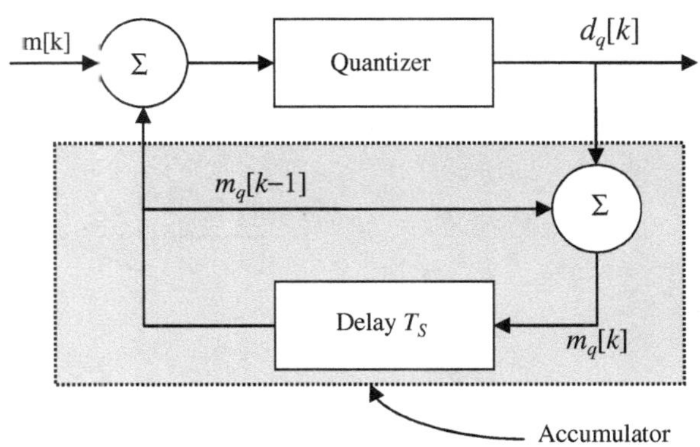

Fig. 2.25 Delta modulation as a special case of DPCM

Here

$$m_q[k] = m_q[k-1] + d_q[k]$$

Hence,

$$m_q[k-1] = m_q[k-2] + d_q[k-1]$$

So,

$$m_q[k] = m_q[k-2] + d_q[k-1] + d_q[k] \tag{2.40}$$

Proceeding iteratively in this manner and considering 0 initial condition, i.e.

$$m_q[0] = 0, \text{ we have } m_q[k] = \sum_{n=0}^{k} d_q[n] \tag{2.41}$$

This shows that the receiver is just an accumulator. If the output $d_q[k]$ is represented by impulses, then the accumulator (receiver) may be realized by an integrator because its output is the sum of the strengths of input impulses (sum of the areas under the impulses). We may replace the feedback portion of the modulator (which is identical to the demodulator) by an integrator. The demodulator output is $m_q[k]$, which when passed through an LPF gives us the retrieved signal. The Fig. 2.26 bellows shows a practical implementation of the delta modulator and demodulator.

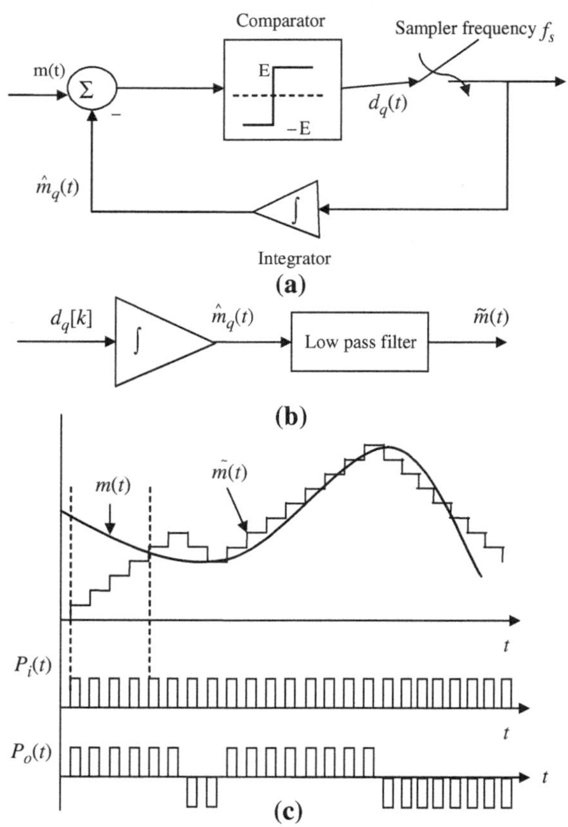

Fig. 2.26 (**a**) Delta modulator, (**b**) Delta demodulator, (**c**) DM waveforms

2.4.1 Drawbacks of Delta Modulation

The waveform $\widetilde{m}(t)$ needs to closely follow the waveform of $m(t)$ in order to the recovered waveform $\widetilde{m}(t)$ resembles $m(t)$. Taking a careful look at the waveform shows situations where $\widetilde{m}(t)$ is unable to follow $m(t)$ as the slope of $m(t)$ is higher than that of $\widetilde{m}(t)$. Also when $m(t)$ is varying slowly, the change of $\widetilde{m}(t)$ is higher enough to cause an error.

2.4.1.1 Slope Overloading

When, at any instant or any duration of time the slope of $m(t)$ is higher then the slope of $\widetilde{m}(t)$, i.e. step size/time period, $\widetilde{m}(t)$ is unable to follow the input message signal $m(t)$. So the recovered waveform will be distorted. Figure 2.27 shows the case.

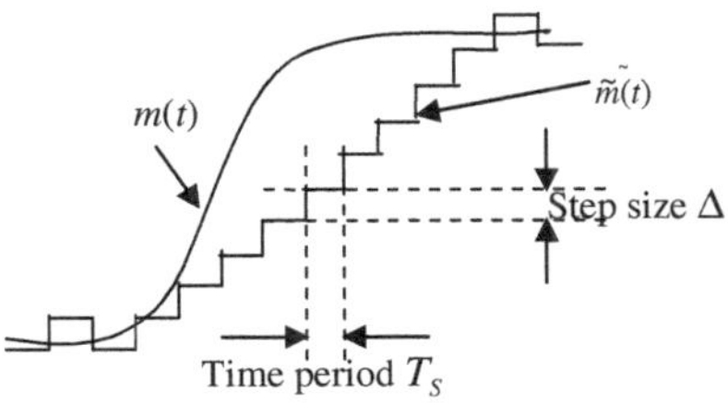

Fig. 2.27 Slope overloading error in delta modulation (DM)

The condition for slope over loading is then,

$$\frac{dm(t)}{dt_{\max}} \geq \frac{\Delta}{T_S} \tag{2.42}$$

Example 2.3 For the signal $s(t) = A \sin 2\pi ft$ calculate the sampling frequency to avoid slope overload error, considering 256 quantization levels and 1 kHz signal frequency.

$$\left.\frac{ds(t)}{dt}\right|_{\max} = A.2\pi f$$

$$\therefore f_s = \frac{1}{T_s} \leq \frac{2A\pi f}{\Delta}$$

$$\Rightarrow f_s \leq M\pi f \qquad \left[\because M = \frac{2A}{\Delta}\right],$$

M = Number of levels,
2A = peak to peak amplitude.

Now, putting M=256, f=1,

$$fs \geq 256 \times 3.14 \times 1\,\text{kHz} = 803.84\,\text{kHz}$$

Therefore, to avoid the error of slope overload, the sampling frequency must be greater than 803.84 kHz.

2.4.1.2 Granular Noise

In other hand, when the input message signal $m(t)$ is varying slowly, the change in $\widetilde{m}(t)$ is comparatively higher that causes error in reconstruction. This type of error or distortion is called Granular noise. Figure 2.28 shows the case.

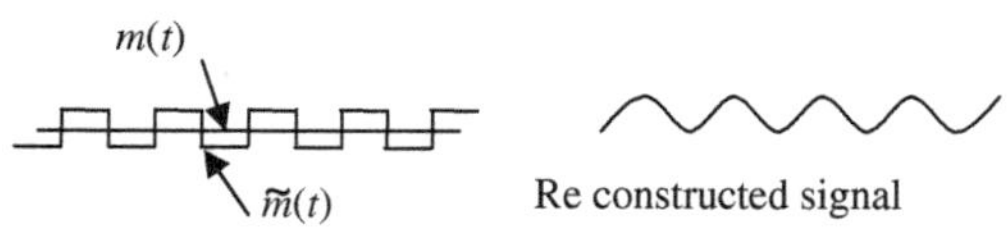

Fig. 2.28 Granular noise error in delta modulation

2.5 Adaptive Delta Modulation

The drawbacks of delta modulation are primarily due to the slope. These drawbacks can be overcome by suitably changing the slop. Slope can be changed either by changing the time period or the step size. Changing the time period will cause in changing the frequency and bandwidth. So it is not possible to vary the time period for a single message. Rather we can vary the step size instead to overcome the limitation. One may question that, by increasing the step size we are increasing the quantization error, because the quantization error is directly proportional to the square of the step size. But here the basic assumption of squared error calculation is missing and does not hold for the present case.

A Delta Modulation system with variable step size is known as the Adaptive Delta Modulation (ADM).

The block diagram of an ADM system is shown in the following Fig. 2.29.

In the concept discussed above, the squaring circuit is responsible for the size of step (Δ) adaptation (tuning). The shaded region shown in the figure above is the 'adaptation algorithm' in generic. Two other well known adaptation algorithms are discussed below.

2.5.1 Song Algorithm

In the year of 1971, C. L. Song et al. [7] proposed a nice algorithm for step size adaptation by which the step size of the predicted waveform ($\widetilde{m}\,(t)$) can nicely be adapted. According to the algorithm, positive slope of the prediction with respect to time results in the next prediction equal to the previous step size added with the

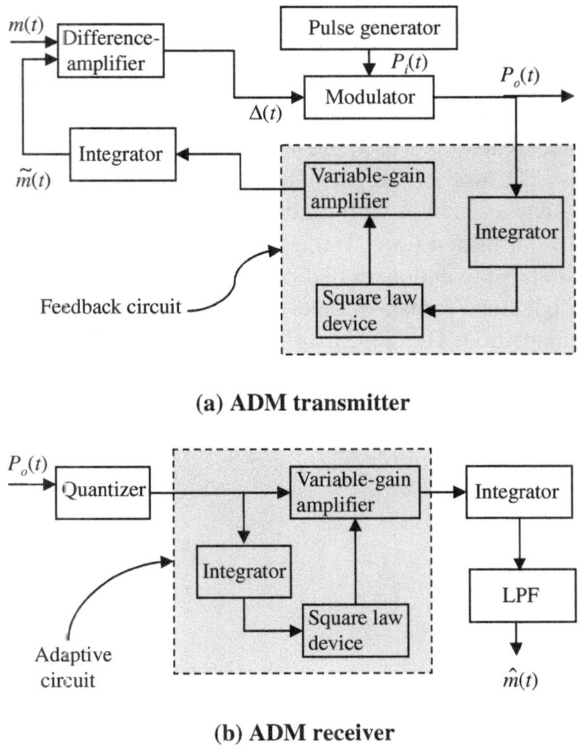

Fig. 2.29 Adaptive delta modulation (ADM) communication system.

step size at 0th prediction(Δ_0), similarly, for negative slope of the prediction with respect to time results in the next prediction equal to the previous step size minus the step size at 0th prediction (Δ_0)

If the nth prediction sample is $\Delta(n)$, the algorithm prescribes the adaptation as shown in the following two equations.

$$\Delta(n) = \Delta(n-1) + \Delta_0 \qquad \text{if} \qquad m(t) > \tilde{m}(t) \tag{2.43}$$

$$\Delta(n) = \Delta(n-1) - \Delta_0 \qquad \text{if} \qquad m(t) < \tilde{m}(t) \tag{2.44}$$

This addition and subtraction is called as *accumulation* in other terms. To combine the Eqs. (2.43) and (2.44) let us assume another parameter called as discrepancy parameter (d_n) as

$$d_n = \begin{cases} +1 & \text{if} \quad m(t) > \tilde{m}(t) \\ -1 & \text{if} \quad m(t) < \tilde{m}(t) \end{cases} \tag{2.45}$$

Now, combining Eqs. (2.43) and (2.44), the nth sample can be calculated using song algorithm of adaptation as

$$\Delta(n) = |\Delta(n-1)\, d_n + \Delta_0 d_{n-1}| \tag{2.46}$$

The song algorithm can now be used for digitization of any type of analog signal shape. Let us take a typical step size analog signal for prediction through accumulation. The ADM waveform through song algorithm is shown below.

The level of decrement is controlled by the new algorithm. In the previous section (adaptation by squaring), the increment and decrement amount is only the step size of 0th instant (Δ_0). But, as shown in the Fig. 2.30, the ADM predicted waveform is showing undue oscillation for typically step like analog signal. For low slope and high slope signals of other types can be tracked more efficiently by this method of adaptation. The slope overload problem is solved totally by the song algorithm. The undue oscillation can be viewed as a special type of granular noise.

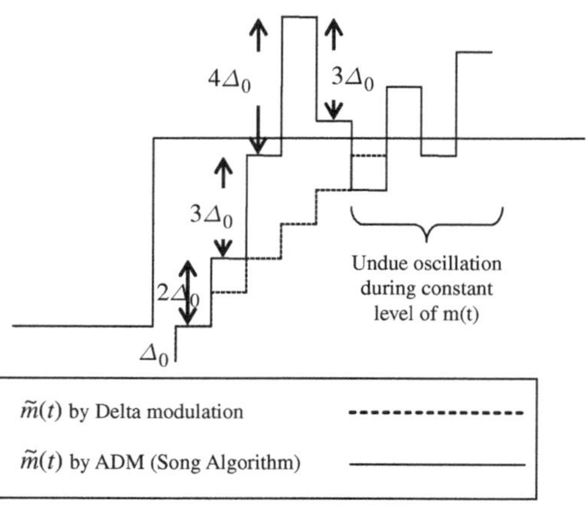

Fig. 2.30 Adaptive delta modulation by song algorithm

2.5.2 *Space-Shuttle Algorithm*

Space-shuttle algorithm is a modification over song algorithm only. We have seen that, the negative slope or maintenance of constant levels of the analog signal cannot be properly tracked by the prediction through adaptive delta modulation by song algorithm. The step size of the accumulated waveform is incremented and decremented by same amount (Δ_0). But, allowance of this linear decrement gives rise to an undue oscillation during constant level of m(t). The new algorithm only changes the rule of decrement. Instead of linear decrement space shuttle algorithm suggests constant decrement of amount Δ_0.

$$\Delta(n) = \Delta(n-1) + \Delta_0 \quad if \quad m(t) > \tilde{m}(t) \tag{2.47}$$

$$\Delta(n) = \Delta_0 \quad if \quad m(t) < \tilde{m}(t) \tag{2.48}$$

Otherwise, the algorithm is very much similar to the song algorithm. Here also, the step size is incremented if the direction of the step at the current clock edge is the same as the previous clock edge. If the direction is opposite, the step size is decremented.

2.6 Sigma-Delta Modulation (SDM)

Delta modulation requires two integrators for modulation and demodulation processes as shown in Fig. 2.31 below. Since integration is a linear operation, the second integrator can be moved before the modulator without altering the overall input/output characteristics. Furthermore, the two integrators in Fig. 2.31a can be combined into a single integrator by the linear operation property (Fig. 2.31b).

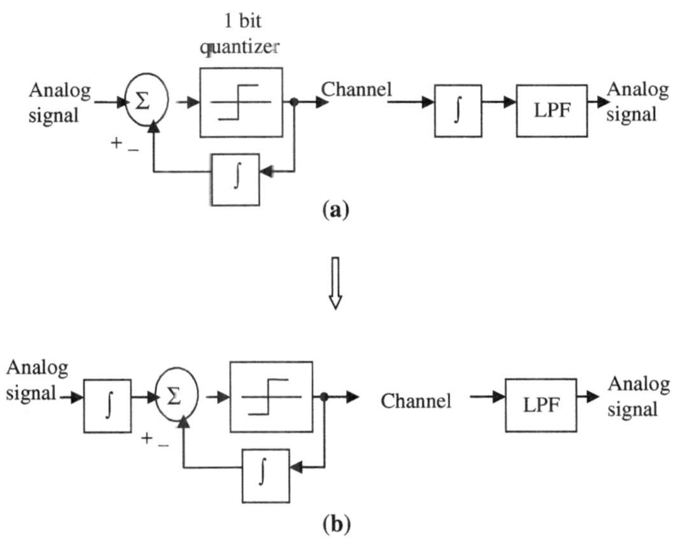

Fig. 2.31 (**a**) Delta modulator and demodulator, (**b**) Modification of (**a**) by integrator matching

The arrangement shown in Fig. 2.32 is the design diagram of a Sigma-Delta (S-D) Modulation system. This structure, besides being simpler, can be considered as being a 'smoothed version' of a 1-bit pulse code modulation (delta modulator). Here, the *smoothness* signifies low pass filtering (i.e., integrated) of the comparator

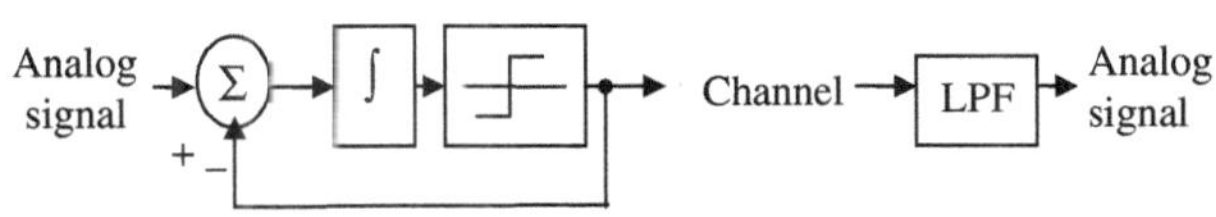

Fig. 2.32 Sigma-Delta modulator with a single integrator

output as shown in Fig. 2.32 and 1 bit PCM signifies a hard limiter of 2 output levels of saturation at the output (Fig. 2.26).

The name Sigma-Delta modulator comes from putting the integrator (sigma) in front of the delta modulator. The use of the integrator in the manner described here has also the following benefits:

1. The low frequency content of the input analog signal is pre-emphasized.
2. Correlation between adjacent samples of the delta modulator input is increased, which improves the overall system performance by reducing the variance of the error signal at the quantizer input.
3. Design of the receiver is simplified.
4. Noise performance is improved as discussed in the following section.

2.6.1 Noise Performance

Sometimes, the S-D modulator is referred to as an interpolative coder. The quantization noise characteristic (noise performance) of such a coder is frequency dependent in contrast to delta modulation. As will be discussed further, this noise-shaping property is well suited to signal processing applications such as digital audio and communication. Like delta modulators, the S-D modulators use a simple course quantizer (comparator). However, unlike delta modulators, these systems encode the integral of the signal itself and thus their performance is insensitive to the rate of change of the signal (Fig. 2.33).

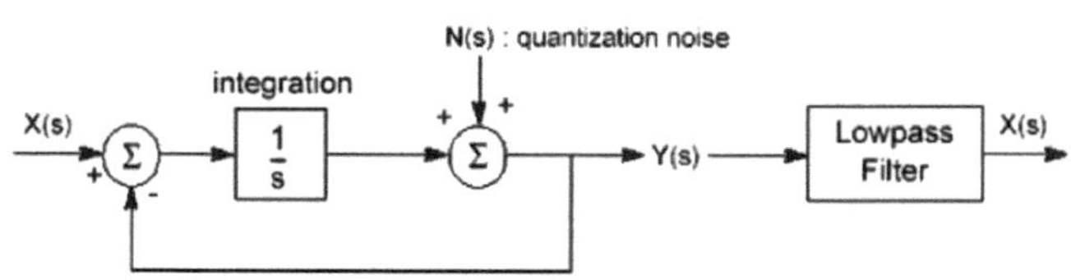

Fig. 2.33 Delta-Sigma modulation in frequency domain representation

- Signal transfer function (when N(S)=0):

$$Y(s) = [X(s) - Y(s)] \times \frac{1}{s} \tag{2.49}$$

$$\frac{Y(s)}{X(s)} = \frac{\frac{1}{s}}{1 + \frac{1}{s}} = \frac{1}{1 + s} \tag{2.50}$$

Clearly, it signifies transfer function of LPF.

- Noise transfer function (when X(S)=0):

$$Y(s) = -Y(s) \times \frac{1}{s} + N(s) \tag{2.51}$$

$$\frac{Y(s)}{N(s)} = \frac{1}{1 + \frac{1}{s}} = \frac{s}{s+1} \tag{2.52}$$

Clearly, it signifies transfer function of HPF.

Transfer function shown in Fig. 2.34 illustrates the modulator's main action. As the loop integrates the error between the sampled signal and the input signal, it low pass filters the signal and high pass filters the noise. In other words, the signal is left unchanged as long as its frequency content doesn't exceed the filter's cutoff frequency, but the S-D loop pushes the noise into a higher frequency band. Grossly over-sampling the input causes the quantization noise to spread over a wide bandwidth and the noise density in the bandwidth of interest (baseband) to significantly decrease.

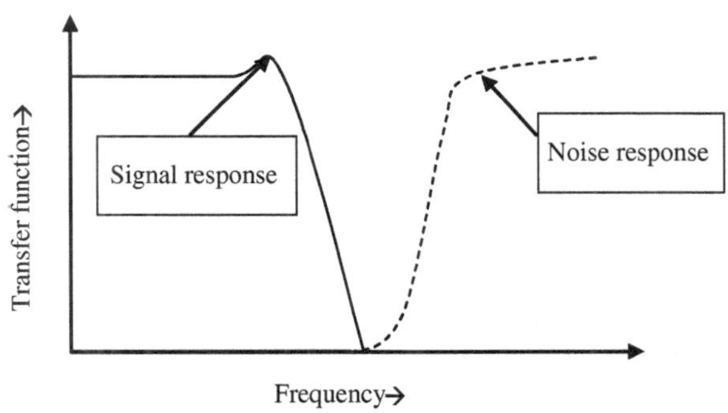

Fig. 2.34 Noise performance of signal delta modulation

2.7 Linear Predictive Coder (LPC)

2.7.1 Concept

From the Sect. 2.3.3 (DPCM prediction), it is understood that, for prediction of the next sample, only the delayed sample may not reduce the error of prediction to our threshold level of satisfaction. For that, we have incorporated a series of delayed samples with suitable scaling factors. The word 'suitable' is important in respect of the design (Fig. 2.35)

The accuracy of the prediction is dependent on the proper choice of the scaling factors. Here the predicted sample is expressible as linear combination of the series of samples as under:

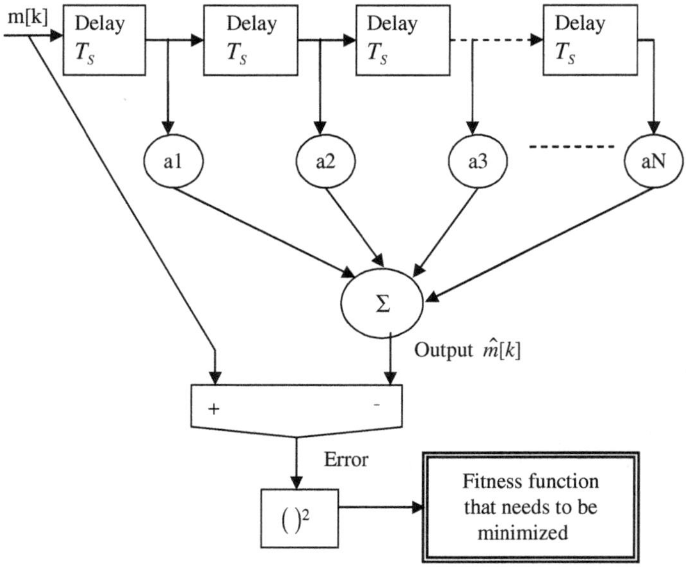

Fig. 2.35 Linear predictive coding with optimization of fitness of prediction

$$
\begin{aligned}
\hat{m}[k] &= a_1 m[k-1] + a_2 m[k-2] + \ldots a_N m[k-N] \\
&= \sum_{i=1}^{N} a_i m[k-i]
\end{aligned}
\tag{2.53}
$$

Next, the error is calculated with respect to real $m[k]$ and the squared value of error (or we can use one squaring circuit to make it sign free) is minimized by any optimization algorithm to find out 'suitable' scaling factors a_i's for proper prediction. This is LSE (least square error) estimation algorithm. Once the scaling factors are fixed for training samples, now our LPC (linear predictive coder) is ready to use for the other query samples also. The optimization process needs to solve a huge number of differential equations to find out proper a_i's for good predictor design. To overcome the hazard of solving the differential equations, we take help of soft computing approaches like Genetic Algorithm (GA), to be discussed in the next section.

2.7.2 Genetic Algorithm Based Approach

Genetic algorithm is a biologically inspired algorithm based on Darwinian evolution, and is a nice algorithm for object optimization. Here, we can parallely discuss about basic GA and application of GA to find out suitable a_i's for minimum squared error.

Genetic algorithms are implemented in a computer simulation in which a population of abstract representations (called chromosomes or the genotype of the genome) of candidate solutions (called individuals, creatures, or phenotypes) to an optimization problem evolves toward better solutions. Traditionally, solutions are represented in binary as strings of 0s and 1s, but other encodings are also possible. Here, we will use real-coded GA. The evolution usually starts from a population of randomly generated individuals and allowed for reproduction. In each generation, the fitness of every individual in the population is evaluated, multiple individuals are stochastically selected from the current population (based on their fitness to the particular problem), and modified (recombined and possibly randomly mutated) to form a new population. The new population is then used in the next iteration of the algorithm. Commonly, the algorithm terminates when either a maximum number of generations has been produced, or a satisfactory fitness level has been reached for the population. If the algorithm has terminated due to a maximum number of generations, a satisfactory solution may or may not have been reached.

A typical genetic algorithm requires:

1. A genetic representation of the solution domain,
2. A fitness function to evaluate the solution domain.

Here, the gene is represented as an array of solutions in real code. The fitness function is defined over the genetic representation and measures the *quality* of the represented solution. The fitness function is always problem dependent. In the present case, the fitness function f is defined as the squared error as given below:

$$f = \varepsilon^2 = \left(m[k] - \hat{m}[k]\right)^2 = \left(m[k] - \sum_{i=1}^{N} a_i m[k-i]\right)^2 \tag{2.54}$$

After having the genetic representation and the fitness function defined, GA proceeds to initialize a population of solutions randomly, and then improve it through repetitive application of mutation, crossover, inversion and selection operators.

Step1: *Population*

A genetic pool is created with 20 chromosomes each with N number of genes with random values, covering the entire range of possible solutions. Here the scaling factors must reside within 0 and 1. Therefore, the initialized random values must be fractions only. Say, the values of genes are the trial solutions of a_i's ($\forall i$) for minimum squared error.

Step2: *Selection*

During each successive generation, a proportion of the existing population is selected to breed a new generation. Individual solutions are selected through a *fitness-based* process. It means, from that 20 strings of arrays (20 chromosomes), the most fit genes will be selected. The values of a_i's ($\forall i$) (i.e., the genes) will be used to find the fitness value according to the Eq. (2) and from those 20 chromosomes (20 trial solutions) 10 better/ fitter will be selected.

Step3: *Reproduction*

The next step is to generate a second generation population of solutions from those selected through genetic operators: crossover (also called recombination), and/or mutation. The Fig. 2.36 clearly shows the process of crossover. The crossover point should be chosen such that number of genes at the left of the cross point of the X chromosome must be exactly equal to the number of genes at the left of the cross point of the Y chromosome.

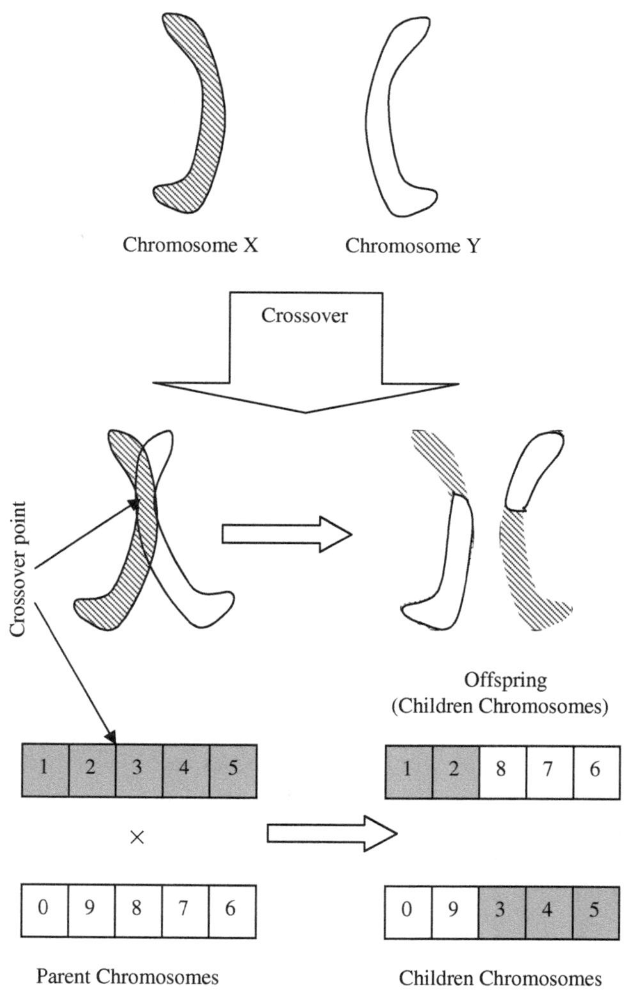

Fig. 2.36 Process of crossover for a 5 variable (N=5) solution

For each new solution to be produced, a pair of 'parent' solutions is selected for breeding from the pool selected previously. By producing a 'child' solution using the above methods of crossover and mutation, a new solution is created which typically shares many of the characteristics of its 'parents'. New parents are selected for each new child, and the process continues until a new population of solutions of appropriate size is generated.

Here we have assumed that, only 5 delays are responsible for the prediction, i.e., N=5. So, two fit chromosomes are chosen as shown in Fig. 2.36. In X chromosome, trial solution for $a_1 = \frac{1}{1+2+3+4+5} = \frac{1}{15}$, $a_2 = \frac{2}{1+2+3+4+5} = \frac{2}{15}$, $a_3 = \frac{3}{1+2+3+4+5} = \frac{3}{15}$ and so on. Similarly, in Y chromosome, trial solution for $a_1 = \frac{0}{0+9+8+7+6} = 0$, $a_2 = \frac{9}{0+9+8+7+6} = \frac{9}{30}$ and so on. After crossover, two new solutions for the scaling factors are created. In the 1st child chromosome, $a_1 = \frac{1}{1+2+8+7+6} = \frac{1}{24}$, $a_2 = \frac{2}{1+2+8+7+6} = \frac{2}{24}$, $a_3 = \frac{8}{1+2+8+7+6} = \frac{8}{24}$ and so on. Similarly, in 2nd child chromosome, solution for $a_1 = \frac{0}{0+9+3+4+5} = \frac{0}{21} = 0$, $a_2 = \frac{9}{0+9+3+4+5} = \frac{9}{21}$ and so on.

The above example shows, the processes ultimately result in the next generation population of chromosomes that is different from the initial generation. Generally the average fitness will have increased by this procedure for the population, since only the best organisms from the first generation are selected for breeding.

Step4: *Termination*

This reproduction and offspring (or child) generation is continuously repeated until a termination condition has been reached. Common terminating conditions are:

- A solution is found that satisfies minimum criteria (threshold of fitness).
- Fixed number of generations reached.
- Allocated budget (computation time/money) reached
- The highest ranking solution's fitness is reaching or has reached
- a plateau such that successive iterations no longer produce better results.
- Manual inspection.
- Combinations of the above.

2.8 MATLAB Programs

2.8.1 *Aliasing*

```
% In CD: ch2_1.m
% Aliasing
% Output: Fig. 2.5
% Programed by Apurba Das (Jan,'10)
```

```
clear all;close all;
freq=100;phase=pi/3;
mag=2;fs=500;Ts=1/fs;
k=1; % number of repetations
num_points=200; % How many points to use, 200 makes it smooth
num_samples=11; % How many samples to simulate reading
                % 11 puts "sampled" under "analog"
step=2/(freq*num_points); % get a nice step size
t=0:step:2*(1/freq);
n=0:num_samples-1;
% x and y are simulated analog functions
x= mag*cos(2*pi*freq*t+phase);
y= mag*cos(2*pi*(freq+k*fs)*t+phase);
% x2 and y2 are simulated sampled version of x and y
x2(n+1)=mag*cos(2*pi*freq*n*Ts+phase);
y2(n+1)=mag*cos(2*pi*(freq+k*fs)*n*Ts+phase);
% Plot the analog signal
subplot(2,1,1);
plot(t,x,'r.-');hold on;plot(t,y,'b-');
title('Simulated analog signals, x=dots y=solid');
xlabel('Time');ylabel('Amplitude');
% Plot "sampled" signals
subplot(2,1,2);
plot(n,x2,'rx');hold on;plot(n,y2,'bo');
title('Simulated Sampled signals, x=x y=o');
xlabel('Time');ylabel('Samples');
figure;
nn=0:0.002:0.02;
plot(t,x,'r.');hold on;plot(t,y,'b-');stem(nn,x2,'kv');
```

References

1. Carlson, A.B., Crilly, P.B., Rutledge, J.C., "Communication Systems – An Introduction to Signals and Noise in Electrical Communication (4th Ed.)", McGraw-Hill International, 2002
2. Lathi, B.P., "Modern Digital and Analog Communication Systems (3rd Ed.)", Oxford University Press, 2005
3. Taub, H., Schilling, D.L., "Principles of Communication Systems", McGraw-Hill, 1986 (Original from the University of Michigan)
4. Smith, B., "Instantaneous Companding of Quantized Signals," *Bell System Technical Journal*, Vol. 36, May 1957, pp. 653–709
5. Kaneko, H., "A Unified Formulation of Segment Companding Laws and Synthesis of Codecs and Digital Compandors," *Bell System Technical Journal*, Vol. 49, September 1970, pp. 1555–1558
6. Skler, B., "Digital Communications: Fundamentals and Applications (2nd Ed.)", Pearson Education, 2001
7. Song, C.L., Schilling, D.L., Garodnick, J. "A Variable Step-Size Robust Delta Modulator", *IEEE Transactions on Communications Technology*, Vol. COM-19, December 1971, pp. 1033–1044

Chapter 3
Digital Baseband Signal Receivers

3.1 Introduction

Baseband signal transmission can directly be defined as digital data transmission without any modulation technique employed. We can say, if raw data is directly been transmitted to the channel intending the receiver, the communication is called as baseband communication. As there is no modulation, there must not be any term as Baseband modulation (as no carrier is present).
High noise immunity of the digital Baseband telecommunication system as shown in the Fig. 3.1, can be assured in 2 ways [3]

i. Improving the structure of the signal being transmitted (as NRZ, RZ, Manchester, etc.)
ii. Developing receiver equipments capable of signal detection in noise (as different types of filters like Matched filter)

Figure 3.2 is showing the entire Baseband receiver or Baseband decoder block. The four configuring blocks are as follows:

1. *Receiving filter*: It recovers the Baseband pulses with best possible signal to noise ratio (SNR). Optimum filter is either Matched filter or Correlator.
2. *Equalizing filter:* This filter is optional in Baseband decoder block. Due to dispersive nature of transmission channel, one effect may come into the received symbols, named as Inter Symbol Interference (ISI). Equalizer is the inverse filter

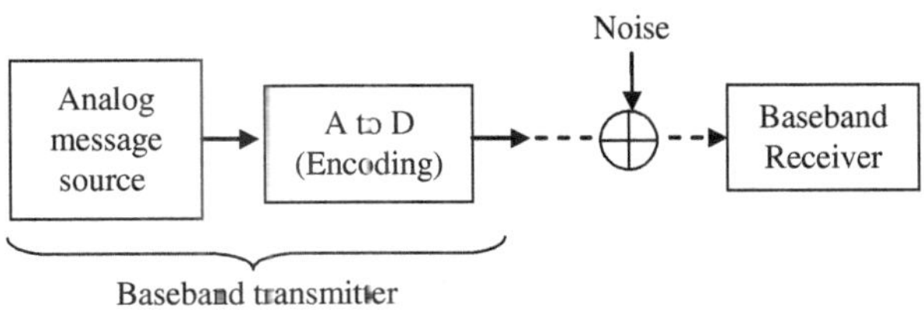

Fig. 3.1 Block diagram of baseband communication system

A. Das, *Digital Communication*, Signals and Communication Technology,
DOI 10.1007/978-3-642-12743-4_3, © Springer-Verlag Berlin Heidelberg 2010

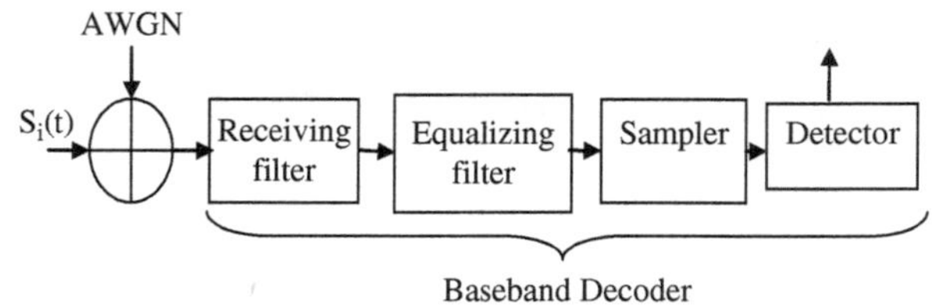

Fig. 3.2 Block diagram of Baseband decoder system

to nullify the effect of ISI. The topic will be discussed in detail in the 5th chapter, equalizer.

3. *Sampler:* It has the detailed timing information of the transmitter, i.e. the block is needed for bit and frame synchronization. At the end of each symbol period T_S, the output of sampler yields a sample $Z(T_S)$, which is a voltage value directly related to transmitted symbol and noise power.
4. *Detector:* In the detection stages, to be precisely said, the main function is to take out the digital logic meaning of the sampled values. The detector is designed on the basis several detection theorem and algorithm.

To be precise, the main constraints of designing the baseband symbol detector are as follows:

a. There must be an appropriate sampling instant at the output side, where the signal strength will be emphasized over the noise strength.
b. Noise figure/SNR must be calculated to get an impression of un-erroneous signal reconstruction.
c. If even the symbol is totally suppressed due to noise, the probability of signal reception must show an appreciable amount.

3.2 Integrate and Dump Type Filter

Let the binary encoded PCM baseband signal be a sequence of voltage levels +V and –V representing logic 1 and logic 0 respectively. In a bit interval, a sample value is taken at a particular time and if it is positive, the decision is that a 1 has been received, and conversely, if it is negative, the decision is that 0 has been received. If at the sampling instant the noise voltage, n(t) is larger the magnitude V, and of opposite polarity, an error will occur. As shown in the Fig. 3.3, at the sampling instant $t = t_1 + \Delta t$, the sampled value is negative (i.e. logic 0), whereas the bit transmitted is logic 1, it is for; the amplitude is higher than V in negative sense.

The probability of error can be reduced if we are able to find the sampling instant at which the sample voltage due to signal is emphasized as against the sample voltage due to noise.

The noise n(t) is assumed to be White, i.e. of power spectral density (PSD) $\eta/2$.

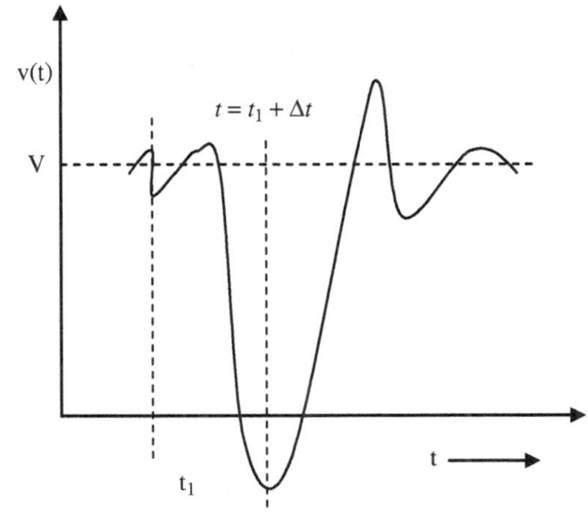

Fig. 3.3 Error due to noise at the sampling instant

The Fig. 3.4 is showing the construction of integrate and dump type filter. The received signal is fed to the inverting terminal of the op-amp. As the feedback path contains one capacitor, therefore it's not difficult to understand, that the circuit is an integrator. Switch S_1 is called as *dump* switch. The switch is being used for abrupt *capacitor discharging*. The switch S_2 *is used for sampling* at the end of uniform time duration, say T. Therefore, it includes a delay of amount T at most.

A current integration circuit includes an operational amplifier having a capacitor connected between its output and inverting input which integrates an input current. To prevent the op amp's output from becoming saturated, a charge dumping circuit dumps a known charge of the opposite polarity to that stored on the capacitor to the op amp's inverting input, thus reducing the charge on the capacitor and preventing the op-amp's output from becoming saturated. A charge dump is triggered whenever the op amp's output exceeds a predetermined trip voltage. Counting the number of charge dumps performed during a given integration period provides a coarse indication of the magnitude of the integrated input current, and the output of the op-amp provides a fine indication.

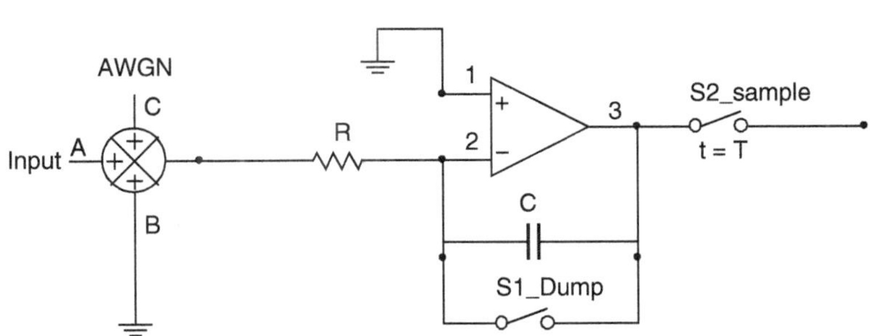

Fig. 3.4 Integrate and dump type filter

Hence the filter or the system can be called as an *Integrate and Dump Type Filter*. The filter is actually combination of *Integration operation* and *Delay operation*.

In the channel, an additive noise, n(t) introduced to the baseband signal s(t), therefore

From the Fig. 3.4, input of the filter

$$v(t) = s(t) + n(t) \tag{3.1}$$

The output of the filter shown in Fig. 3.4, is

$$\begin{aligned} v_o(T) &= \frac{1}{\tau}\int_0^T \{s(t) + n(t)\}dt; \text{ where, } \tau = \text{time constant} = \text{RC} \\ &= \frac{1}{\tau}\int_0^T s(t)dt + \frac{1}{\tau}\int_0^T n(t)dt \end{aligned} \tag{3.2}$$

As shown in the Fig. 3.5, sample voltage due to signal

$$\begin{aligned} s_o(T) &= \frac{1}{\tau}\int_0^T s(t)dt \\ &= \frac{1}{\tau}\int_0^T Vdt = \frac{VT}{\tau} \end{aligned} \tag{3.3}$$

Sample voltage due to noise

$$n_o(T) = \frac{1}{\tau}\int_0^T n(t)dt \tag{3.4}$$

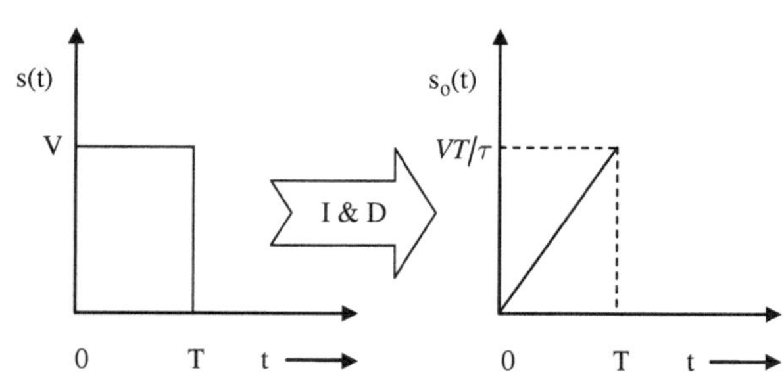

Fig. 3.5 Input and output signal of an integrator

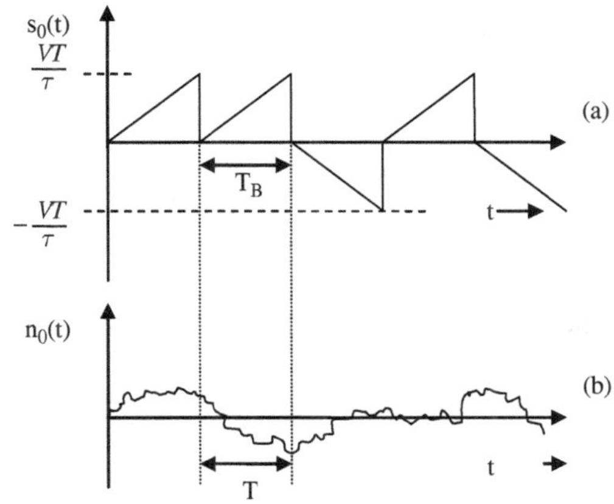

Fig. 3.6 Detection of sampling instant. (**a**) Signal output of integrate and dump filter with input 11010. (**b**) Noise output of integrate and dump filter

The filter output $s_o(t)$ is a ramp voltage, as integration of step is obviously a ramp signal, as shown in Fig. 3.5. At the end of bit interval (T), the magnitude of $s_o(t)$, i.e. $s_o(T)$ is VT/τ or $-VT/\tau$, wherever applicable, for logic 1 and 0 respectively, whereas, $n_o(t)$ has random values at each bit interval.

Thus, at the end of each bit interval, the sample voltage due to signal is emphasized as compared to the sample voltage due to noise. Hence, the end of bit interval is selected as sampling instant (Fig. 3.6).

3.2.1 Noise Power and Variance

From the accuracy point of view, noise power is an important area of concern. If we consider, the noise to be Additive White Gaussian Noise (AWGN), with power spectral density (PSD) $G_n(f)$, the integrate and dump type filter processes the noise PSD and result the output PSD $G_{no}(f)$ (Fig. 3.7). If we consider the transfer function of the filter to be H(f), the relation between $G_n(f)$ and $G_{no}(f)$ would be

$$G_{no}(f) = G_n(f)\,|H_n(f)|^2 \tag{3.5}$$

Here, rms noise power $N_0 = \overline{n_o^2(T)} = \sigma_o^2 =$ variance

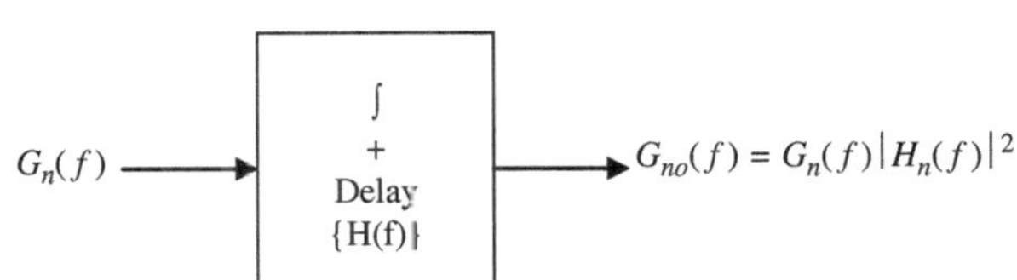

Fig. 3.7 Input and output noise PSD relation

For integration, in the block H(f), the sign $\int$ is replaced by $\frac{1}{s}$ in frequency domain. If the gain of the system is unity, for integration, transfer function is

$$\therefore T.F. = \frac{1}{s} = \frac{1}{j\omega} = \frac{1}{j2\pi f} \tag{3.6}$$

And, for delay

$$\begin{aligned} &\text{if, } x(t) \overset{F.T}{\longleftrightarrow} X(f) \\ &\text{then, } x(t-t_0) \overset{F.T}{\longleftrightarrow} e^{-j\omega t_0} X(f) \end{aligned} \tag{3.7}$$

Combining Eqs. (3.6) and (3.7), and also including the time constant term, τ for the constant forward path gain factor, we get, the transfer function for the integrate and dump type filter as

$$\begin{aligned} H(f) &= \frac{1}{j\omega\tau}\left[1 - e^{-j\omega T}\right] \\ &= \frac{1}{j\omega\tau}\left[1 - \cos\omega T + j\sin\omega T\right] \end{aligned} \tag{3.8}$$

[Combined using –ve sign, as the input is given to the inverting terminal of the op-amp]

From Eq. (3.8),

$$\begin{aligned} |H(f)|^2 &= \frac{1}{(\omega\tau)^2}\left[(1-\cos\omega T)^2 + \sin^2\omega T\right] \\ &= \frac{1+1-2\cos\omega T}{(\omega\tau)^2} \\ &= 2 \times \frac{2\sin^2\frac{\omega T}{2}}{(\omega\tau)^2} \\ &= \frac{\sin^2\pi f T}{(\pi f\tau)^2} \\ &= \left(\frac{T}{\tau}\right)^2 \times \frac{\sin^2\pi f T}{(\pi f T)^2} \end{aligned} \tag{3.9}$$

Now, as the noise is White, the PSD of input noise {G(f)}is $\eta/2$.

Therefore, noise output power,

$$N_0 = \int_{-\infty}^{\infty} G_{no}(f)df$$

$$= \int_{-\infty}^{\infty} G_n(f)\left(\frac{T}{\tau}\right)^2 \times \frac{\sin^2 \pi fT}{(\pi fT)^2}df$$

From definition of PSD and Eq. (3.5). Putting $G_n(f) = \eta/2$, we get

$$N_o = \frac{\eta}{2} \times \left(\frac{T}{\tau}\right)^2 \times \frac{1}{\pi T}\int_{-\infty}^{\infty}\left(\frac{\sin \pi fT}{\pi fT}\right)^2 d(\pi fT) \qquad \left[\because \int_{-\infty}^{\infty}\frac{\sin^2 x}{x^2}dx = \pi\right]$$

$$= \frac{\eta T}{2\tau^2} \tag{3.10}$$

3.2.2 *Figure of Merit*

From Eqs. (3.3) and (3.10), the signal power and noise power of the detected symbol can be derived as follows.

$$\text{Signal power} = P_s = [s_o(T)]^2 = \left(\frac{VT}{\tau}\right)^2$$

$$\text{Noise power} = P_n = \frac{\eta T}{2\tau^2}$$

Therefore, the figure of merit of the receiver is the SNR (Signal to Noise Ratio), given by

$$SNR = \frac{P_S}{P_N} = \frac{\left(\frac{VT}{\tau}\right)^2}{\frac{\eta T}{2\tau^2}} = \frac{2V^2T}{\eta} \tag{3.11}$$

3.2.3 *Probability of Error*

The probability density of the noise sample $n_o(T)$ is Gaussian as shown in figure. Hence,

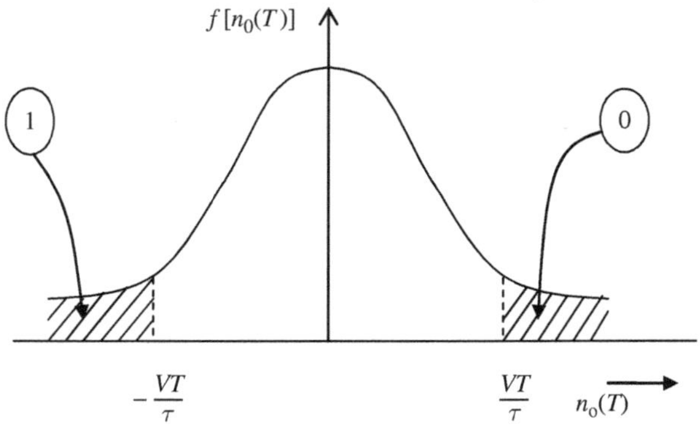

Fig. 3.8 PDF of noise sample

$$f\,[n_o(T)] = \frac{e^{-n_o^2(T)/2\sigma_0^2}}{\sqrt{2\pi\sigma_0^2}} \tag{3.12}$$

Since, at the end of bit interval, $S_0\,(T) = -VT/\tau$ (for logic 0 representation), the error will occur if $n_0\,(T) > VT/\tau$. Hence, the probability of error in this case is given by the marked area (0) shown in Fig. 3.8.

Hence,

$$\begin{aligned} P_e &= \int_{\frac{VT}{\tau}}^{\infty} f\,[n_o(T)]dn_o(T) \\ &= \int_{\frac{VT}{\tau}}^{\infty} \frac{e^{-n_o^2(T)/2\sigma_0^2}}{\sqrt{2\pi\sigma_0^2}} dn_o(T) \end{aligned} \tag{3.13}$$

Substituting $x = n_o(T)/\sqrt{2}\sigma_0$ and putting the value of σ_0^2, we get

$$\begin{aligned} P_e &= \frac{1}{2} \times \frac{2}{\pi} \int_{V\sqrt{\frac{T}{\eta}}}^{\infty} e^{-x^2} dx \\ &= \frac{1}{2} \times erfc\left(V\sqrt{\frac{T}{\eta}}\right) \\ &= \frac{1}{2} \times erfc\left(\frac{V^2T}{\eta}\right)^{\frac{1}{2}} \\ &= \frac{1}{2} \times erfc\left(\frac{E_S}{\eta}\right)^{\frac{1}{2}} \end{aligned} \tag{3.14}$$

where, Signal Energy $= E_S = V^2T$

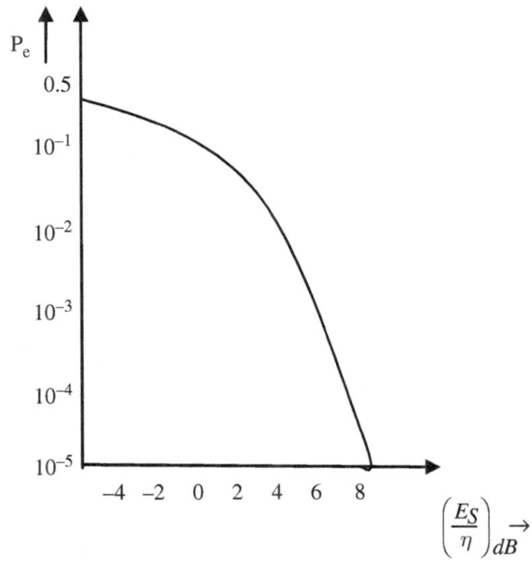

Fig. 3.9 Probability of error versus E_s/η

Also, at the end of the bit interval, $S_0\,(T) = VT/\tau$ for logic 1, the error will occur if $n_0\,(T) < -VT/\tau$. Hence, the probability of error in this case is given by the marked area (1) shown in Fig. 3.9. Because of the symmetry of the Gaussian curve, both the marked areas (0) and (1) are of equal area, and so the probability of error defined by the Eq. (3.14) is sufficient to deduce the entire error probability.

The graph of P_e versus E_S/η in db is a clear representation of error probability improvement with respect to signal strength [2]. The maximum value of P_e is ½ which implies, even if the signal is entirely lost in the noise, the receiver cannot detect wrong more than ½ the chance on the average.

3.3 The Optimum Filter

As discussed in the previous section, the probability of error is the characterizing parameter of the filter. If one filter provides minimum probability of error, the filter or integrator can be called as an *optimum filter.*

The above Fig. 3.10 is showing a generalized receiver for binary coded PCM. If the input is $S_0(t)$, the output is $v_0(T) = S_{00}(T) + n_0(T)$ and if, the input is $S_1(t)$, the output is $v_0(T) = S_{01}(T) + n_0(T)$. Therefore is no noise is present there, the output of the entire receiver assembly would be $S_{00}(T)$ or $S_{01}(T)$. Thus, the presence of noise is ensured by the closeness of $n_0(T)$ to $S_{00}(T)$ or $S_{01}(T)$. The decision boundary is, therefore, the average of $S_{00}(T)$ and $S_{01}(T)$, i.e. $\dfrac{S_{00}(T) + S_{01}(T)}{2}$.

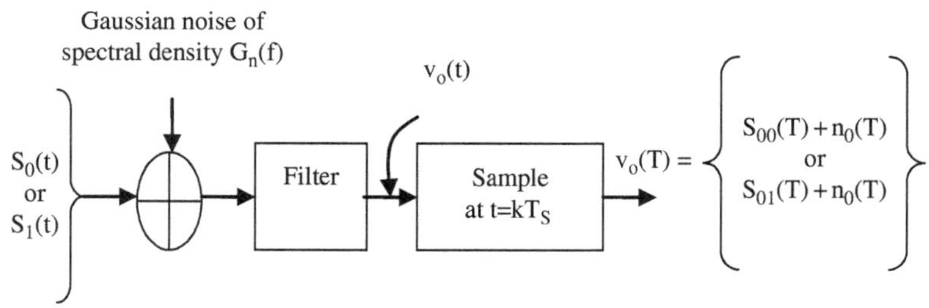

Fig. 3.10 Generalized receiver for PCM

A case has been considered where $S_{00}\,(T) > S_{01}\,(T)$, and $S_1\,(T)$ is transmitted. It is obvious that, an error will occur if at sampling instant, the noise $n_0\,(T)$ is larger than $\frac{1}{2}\,[S_{00}(T) + S_{01}(T)] - S_{01}(T)$ or $\frac{S_{00}(T) - S_{01}(T)}{2}$.

Thus the error will occur when

$$n_0\,(T) \geq \frac{S_{00}(T) - S_{01}(T)}{2} \tag{3.15}$$

Hence, the probability of error is

$$P_e = \int\limits_{\frac{S_{00}(T)-S_{01}(T)}{2}}^{\infty} \frac{e^{\frac{n_0^2(T)}{2\sigma_0^2}}}{\sqrt{2\pi\sigma_0^2}} dn_0(T)$$

$$\Rightarrow P_e = \frac{1}{2}\frac{2}{\sqrt{\pi}} \int\limits_{\frac{S_{00}(T)-S_{01}(T)}{2\sqrt{2}\sigma_0}}^{\infty} e^{-x^2} dn_0(T), \text{ assuming } x = n_0\,(T)/\sqrt{2}\sigma_0$$

$$\Rightarrow P_e = \frac{1}{2} erfc\left[\frac{S_{00}\,(T) - S_{01}\,(T)}{2\sqrt{2}\sigma_0}\right] \tag{3.16}$$

Equation (3.16) is nothing but the generalized form of the Eq. (3.14).

Optimum filter is the filter which optimizes the ratio

$$\gamma = \left[\frac{S_{00}\,(T) - S_{01}\,(T)}{\sigma_0}\right] \tag{3.17}$$

It is to be noted that, the optimization parameter γ is dependent on the difference signal not the individual. The reason is as follows.

When we consider a given signal's of Logic 1 and Logic 0 then we find that whatever the format of the modulated digital signal may be there is always a relation between them. During probability of error calculation we find that the noise pulls down the higher level appreciably and pulls up/boosts the lower level appreciably, but we must keep in mind that both distorted levels should not cross each other. If the distance between two logic levels is high then probability of error occurrence is low.

The difference between the transmitted symbols and difference between the output symbols of the filter, $p_i(t)$ and $p_o(t)$ respectively, are given by,

$$p_i(t) = S_1(t) - S_0(t) \tag{3.18}$$

$$p_o(t) = S_{01}(t) - S_{00}(t) \tag{3.19}$$

If H(f) is the transfer function of the filter,

$$P_o(f) = H(f)\, P_i(f) \tag{3.20}$$

where, $P_i(f)$ and $P_0(f)$ are the Fourier transform of $p_i(t)$ and $p_o(t)$ respectively. Now taking Inverse Fourier Transform, the sampled filtered output difference can be obtained as

$$p_0(T) = \int_{-\infty}^{\infty} P_0(f) e^{j2\pi fT} df = \int_{-\infty}^{\infty} H(f) P_i(f) e^{j2\pi fT} df \tag{3.21}$$

From Parseval's Theorem and Eq. (3.5), the normalized noise power, i.e., noise variance is

$$\sigma_0^2 = \int_{-\infty}^{\infty} G_{no}(f) df = \int_{-\infty}^{\infty} |H_n(f)|^2 G_n(f) df \tag{3.22}$$

Now, from Eqs. (3.17), (3.21) and (3.22),

$$\gamma^2 = \frac{p_0^2(T)}{\sigma_0^2} = \frac{\left| \int_{-\infty}^{\infty} H(f)\, P_i(f)\, e^{j2\pi fT} df \right|^2}{\int_{-\infty}^{\infty} |H_n(f)|^2 G_n(f) df} \tag{3.23}$$

To proceed further with this equation, the help of 'Schwartz inequality' is to be taken. It states that,

$$\left|\int_{-\infty}^{\infty} X(f)\,Y(f)\,df\right|^2 \le \int_{-\infty}^{\infty} |X(f)|^2 df \times \int_{-\infty}^{\infty} |Y(f)|^2 df$$
$$\text{i.e.,}\quad \frac{\left|\int_{-\infty}^{\infty} X(f)\,Y(f)\,df\right|^2}{\int_{-\infty}^{\infty} |X(f)|^2 df} \le \int_{-\infty}^{\infty} |Y(f)|^2 df \tag{3.24}$$

The equality holds if

$$X(f) = kY^*(f) \tag{3.25}$$

where, k is an arbitrary constant and $Y^*(f)$ is the complex conjugate of $Y(f)$. Now, comparing Eqs. (3.23) and (3.24),

$$X(f) = \sqrt{G_n(f)} H_n(f) \tag{3.26}$$

$$\text{and } Y(f) = \frac{P_i(f)\, e^{j2\pi fT}}{\sqrt{G_n(f)}} \tag{3.27}$$

From Eqs. (3.23) and (3.24),

$$\gamma^2 \le \int_{-\infty}^{\infty} \frac{|P_i(f)|^2}{G_n(f)} df \tag{3.28}$$

γ^2 holds the maximum value when (from Eq. 3.25)

$$\sqrt{G_n(f)} H_n(f) = k \times \left[\frac{P_i(f)\, e^{j2\pi fT}}{\sqrt{G_n(f)}}\right]^*$$
$$\Rightarrow H(f) = k\frac{P_i^*(f)}{G_n(f)} e^{-j2\pi fT} \tag{3.29}$$

and if the condition depicted in Eq. (3.29) is satisfied, the parameter[1] is optimized as

$$\gamma_{\max}^2 = \int_{-\infty}^{\infty} \frac{|P_i(f)|^2}{G_n(f)} df \tag{3.30}$$

[1]For mathematical convenience, γ^2 is maximized instead of γ.

3.4 The Matched Filter

If a signal s(t) gets corrupted by AWGN, the optimum filter with impulse response matched to s(t) maximizes the output SNR. This optimum filter is therefore called as *Matched Filter.* PSD (Power Spectral density) of white noise if $G_n(f) = \eta/2$. Rewriting the Eq. (3.29),

$$H(f) = k\frac{P_i^*(f)}{\eta/2}e^{-j2\pi fT} \tag{3.31}$$

3.4.1 Impulse Response

The impulse response of the filter is

$$\begin{aligned} h(t) &= F^{-1}\left[H(f)\right] \\ &= \frac{2k}{\eta}\int_{-\infty}^{\infty} P_i^*(f)e^{-j2\pi fT}e^{j2\pi ft}df \\ &= \frac{2k}{\eta}\int_{-\infty}^{\infty} P_i^*(f)e^{-j2\pi f(t-T)}df \\ &= \frac{2k}{\eta}\int_{-\infty}^{\infty} P_i(f)e^{-j2\pi f(t-T)}df \end{aligned}$$

$$\left[\because \int_{-\infty}^{\infty} P_i^*(f)df = \int_{-\infty}^{\infty} P_i(f)df,\ as\ h^*(t) = h(t) \text{ for physically realizable filters}\right]$$

$$\begin{aligned} \therefore h(t) &= \frac{2k}{\eta}p_i\,(T-t) \\ &= \frac{2k}{\eta}\{S_1\,(T-t) - S_0\,(T-t)\} \end{aligned} \tag{3.32}$$

3.4.2 Probability of Error

From Parseval's theorem,

$$\int_{-\infty}^{\infty} |P(f)|^2 df = \int_{-\infty}^{\infty} p^2(t)dt = \int_{0}^{T_B} p^2(t)dt \tag{3.33}$$

as *p(t)* exists only for time 0 to T_B.

Now, from the expression of optimizing parameter of optimum filter (Eq. 3.30),

$$\begin{aligned}\gamma_{\max}^2 &= \frac{2}{\eta}\int_{-\infty}^{\infty}\{S_1(t)-S_0(t)\}^2\,dt \\ &= \frac{2}{\eta}\int_{-\infty}^{\infty}\left\{S_1^2(t)+S_0^2(t)-2S_1(t)S_0(t)\right\}dt \\ &= \frac{2}{\eta}\left[E_{S1}+E_{S0}-2E_{S1S0}\right]\end{aligned} \tag{3.34}$$

where, E_{S1}, E_{S0} are the energies of S_1(t) and S_0(t) respectively, and E_{S1S0} is the joint energy of the product waveform $\sqrt{S_1(t)S_0(t)}$.

When $S_1(t) = -S_0(t)$, $E_{S1} = E_{S0} = -E_{S1S0} = E_S$. From Eq. 3.34,

$$\gamma^2{}_{\max} = \frac{8E_S}{\eta} \tag{3.35}$$

With the value of optimizing parameter in Eq. (3.35), probability of error from Eqs. (3.16) and (3.19), is given by

$$\begin{aligned}P_e &= \frac{1}{2}erfc\left[\frac{p_o(T)}{2\sqrt{2}\sigma_0}\right] \\ &= \frac{1}{2}erfc\left[\frac{p_o^2(T)}{8\sigma_0^2}\right]^{1/2} \\ &= \frac{1}{2}erfc\left[\frac{\gamma^2}{8}\right]^{1/2}\end{aligned} \tag{3.36}$$

From the basic concept of optimal filter, P_e is minimum when $\gamma^2 = \gamma^2{}_{\max}$.

Therefore,

$$\begin{aligned}P_e|_{Min} &= \frac{1}{2}erfc\left[\frac{\gamma^2{}_{Max}}{8}\right]^{1/2} \\ P_e|_{Min} &= \frac{1}{2}erfc\left[\frac{E_S}{\eta}\right]^{1/2}\end{aligned} \tag{3.37}$$

Equation 3.37 signifies that *the error probability depends only on the signal energy and not on the signal wave shape.*

Example 3.1 A binary baseband digital communication system employs the signal

$$s(t) = \begin{cases} \dfrac{1}{\sqrt{T_S}}; 0 \le t \le T_S \\ 0; \text{otherwise} \end{cases}$$

Determine the output of the matched filter of this signal.

To get the matched filter response (output) of the signal, the input signal is to be convolved with the impulse response of the matched filter.
From Eq. 3.32, the impulse response is

$$h(t) = S\left(T_S - t\right),$$

with the filter constant part=1 and symbol for binary digit zero = 0.

$$y(t) = \int_{-\infty}^{\infty} s\left(\lambda\right) h\left(t - \lambda\right) d\lambda = \int_{-\infty}^{\infty} s\left(\lambda\right) s\left\{T_S - \left(t - \lambda\right)\right\} d\lambda$$

$$= \int_{-\infty}^{\infty} s\left(\lambda\right) s\left\{\lambda - \left(t - T_S\right)\right\} d\lambda$$

As shown in Fig. 3.11, the time reversed and delayed s(t) extends from $(t - T_S)$ to t. Out of this range, the integrand is nonzero for 0 to t on the region $t < T_S$ and for $(t - T_S)$ to T_S in the region $t > T_S$

$$y(t) = \begin{cases} \int_0^t s\left(\lambda\right) s\left\{\lambda - \left(t - T_S\right)\right\} d\lambda; t<T_S \\ \int_{t=T_S}^{T_S} s\left(\lambda\right) s\left\{\lambda - \left(t - T_S\right)\right\} d\lambda; t>T_S \end{cases}$$

$$= \begin{cases} \int_0^t \dfrac{1}{\sqrt{T_S}} \dfrac{1}{\sqrt{T_S}} d\lambda; t<T_S \\ \int_{t=T_S}^{T_S} \dfrac{1}{\sqrt{T_S}} \dfrac{1}{\sqrt{T_S}} d\lambda; t>T_S \end{cases}$$

$$= \begin{cases} \dfrac{1}{T_S} \left[t - 0\right]; t<T_S \\ \dfrac{1}{T_S} \left[T_S - \left\{t - T_S\right\}\right]; t>T_S \end{cases}$$

$$= \begin{cases} \dfrac{t}{T_S}; t<T_S \\ \dfrac{2T_S - t}{T_S}; t>T_S \end{cases}$$

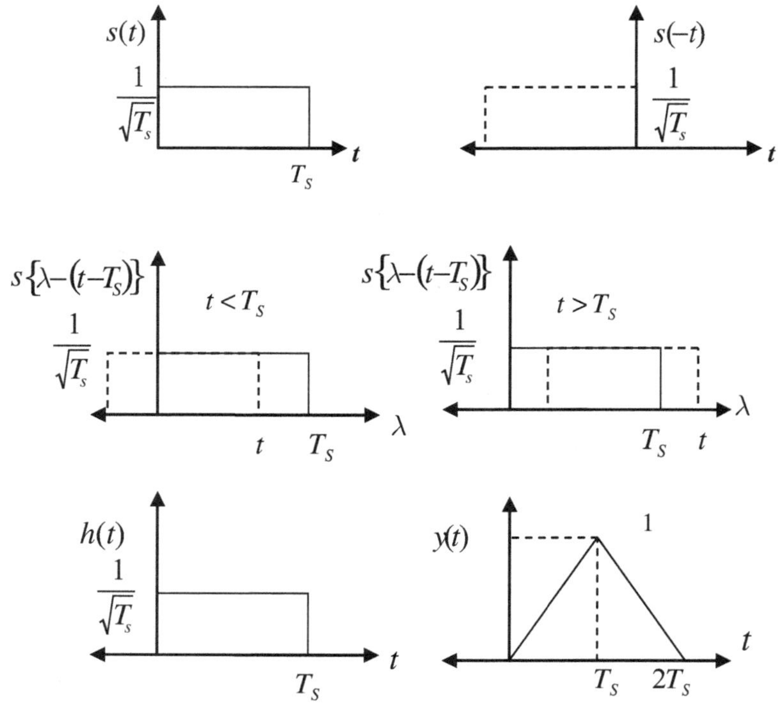

Fig. 3.11 Matched for a rectangular pulse

This is a double slope triangular waveform as shown in the Fig. 3.11. At $t = T_S$, the output sample is 1.

Therefore, in general, the response y(t) of a matched filter to the signal s(t), is

$$y(t) = \int_0^t s(\tau)s(T_S - t + \tau)\,d\tau \tag{3.38}$$

which is the autocorrelation function of the signal s(t).

3.4.3 Properties of Matched Filter

In the operations and principles, matched filters acts on the basis of some unique properties [3].

Property 3.1 If a signal s(t) gets corrupted by AWGN, the filter with impulse response matched to s(t) maximizes the output SNR.

This property can be justified considering a received signal r(t) consisting of a signal s(t) and a zero-mean Gaussian noise n(t) with power spectral density $\eta/2$. If r(t) is passed through a filter (system) with impulse response h(t), then the filter output sampled as T_S can be written as,

$$
\begin{aligned}
y(t) &= \int_{0}^{t=T_S} r(\tau)\, h(t-\tau) d\tau \\
&= \int_{0}^{T_S} s(\tau)\, h(T_S-\tau) d\tau + \int_{0}^{T_S} n(\tau)\, h(T_S-\tau) d\tau \\
&= y_S(T_S) + y_n(T_S)
\end{aligned}
\tag{3.39}
$$

where, $y_S(T_S)$ and $y_n(T_S)$ are the signal and noise component of the filter output. The problem is to select a filter impulse that maximizes the output SNR defined as,

$$
SNR = \frac{y_S^2(T_S)}{E_S\left[y_n^2(T_S)\right]} \tag{3.40}
$$

Now,

$$
\begin{aligned}
E\left[y_n^2(T_S)\right] &= \int_{0}^{T_S}\int_{0}^{T_S} E\left[n(\tau)n(t)\right] h(T_S-\tau)h(T_S-t) dt d\tau \\
&= \frac{\eta}{2}\int_{0}^{T_S}\int_{0}^{T_S} \delta(t-\tau)h(T_S-\tau)h(T_S-t) dt d\tau \\
&= \frac{\eta}{2}\int_{0}^{T_S} h^2(T_S-t) dt
\end{aligned}
\tag{3.41}
$$

Therefore, putting the value of noise variance from Eq. (3.41) into Eq. (3.40),

$$
\begin{aligned}
SNR &= \frac{\left[\int_{0}^{T_S} s(\tau)\, h(T_S-\tau) d\tau\right]^2}{\frac{\eta}{2}\int_{0}^{T_S} h^2(T_S-t) dt} \\
&= \frac{\left[\int_{0}^{T_S} h(\tau)\, s(t-\tau) d\tau\right]^2}{\frac{\eta}{2}\int_{0}^{T_S} h^2(T_S-t) dt}
\end{aligned}
$$

From Schwartz inequality discussed in the previous sections, maximum value of SNR can be obtained as,

$$SNR_{\max} = \frac{2}{\eta}\int_{0}^{T_S} s^2(t)dt = \frac{2E_S}{\eta} \tag{3.42}$$

Property 3.2 The impulse response of the matched filter is the mirror image of the signal.

From Schwartz inequality it is easily understandable that, the impulse response of the matched filter is

$$h(t) = \begin{cases} k \times s(T_S - t)\,; 0 \le t \le T_S \\ 0; elsewhere \end{cases} \tag{3.43}$$

This implies that the impulse response of the filter that is responsible for maximum signal to noise ratio, is the mirror image of the message signal, delayed by the symbol duration.

Property 3.3 Matched filter does not contribute inter symbol interference (ISI).

For the delay T_S and the convolution operation, the match filer output is always non-zero for $t > 2T_S$. Hence, match filter, though introduces a delay, does not introduce ISI as it makes the output to go to zero at the next sampling instant.

3.5 The Correlator

The correlator is a *coherent* baseband receiver system. The term 'correlation' means similarity. The present system takes the noise accumulated signal from the communication channel and generates a known difference waveform with respect to raw signal component, locally. The similarity of all the components on the received signal is then measured with the local signal. Obviously the similarity with the signal component is much higher than the similarity (correlation) with the noise component. As the system maximizes the correlation, the system is called as *correlator.*

The in-system arrangement is shown in the Fig. 3.12. The received signal along-with noise is multiplied by a locally generated signal $S_1(t) - S_0(t)$. The output is then passed through an integrator (LPF) whose response is samples at $t = T_S$. Immediately after each sampling, the energy storing element of the integrator is discharged.

The output signal and noise are

$$S_o(T_S) = \frac{1}{\tau}\int_{0}^{T_S} S_i(t)\,[S_1(t) - S_0(t)]\,dt \tag{3.44}$$

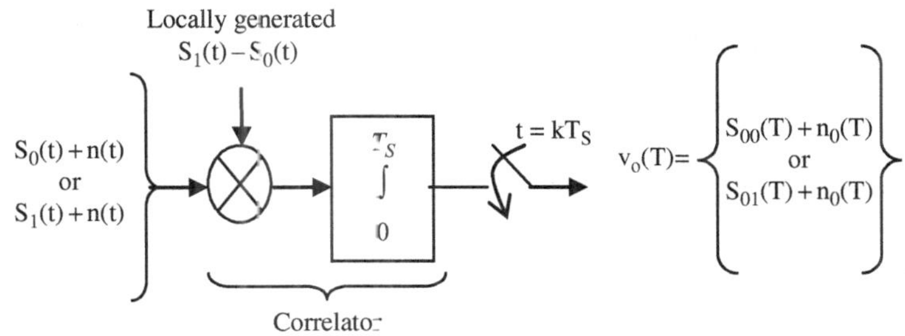

Fig. 3.12 Coherent system of baseband signal reception

$$n_o(T_S) = \frac{1}{\tau}\int_0^{T_S} n(t)\,[S_1(t) - S_0(t)]\,dt \tag{3.45}$$

where, $S_i(t)=S_1(t)$ or $S_0(t)$.

It is obvious that, from Eqs. (3.44) and (3.45), at the output of the correlator, the signal part will dominate the noise as because, the product modulator performs for all possible inputs and chooses the best out of them. The signal part will result in maximum correlation (Fig. 3.13). Therefore, the SNR will be higher.

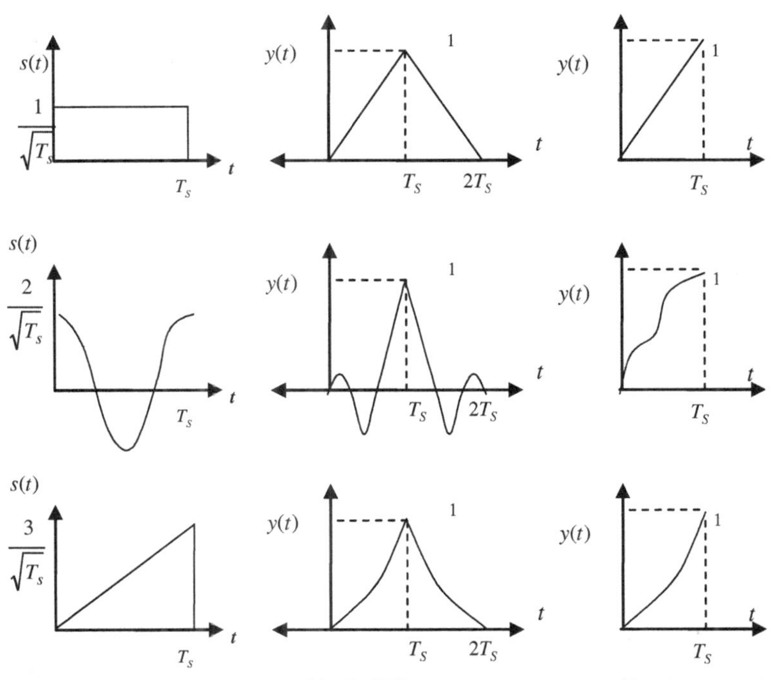

Fig. 3.13 Comparison between correlator and matched filter

3.6 Simulink Communication Block Set Example

Integrate and Dump

Integrate, resetting to zero periodically and reducing by a modulus

Library

Integrators, in Basic Communication Functions [1]

Description

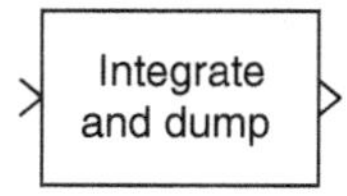

Fig. 3.14 Integrate and dump filter in simulink based model file

The Integrate and Dump block integrates the input signal in discrete time, resets to zero according to a fixed schedule, and reduces modulo the *Absolute value bound* parameter. If the *Absolute value bound* parameter is K, then the block output is strictly between −K and K.

The reset times are the positive integral multiples of the *Integration period* parameter. At each reset time, the block performs its final integration step, sends the result to the output port, and then clears its internal state for the next time step.

The input can be a scalar, a sample-based vector, or a frame-based row vector. The block processes each vector element independently.

This block uses the Forward Euler integration method.

Dialog Box

Fig. 3.15 Parameter adjustment of the Integrate and dump filter block in simulink

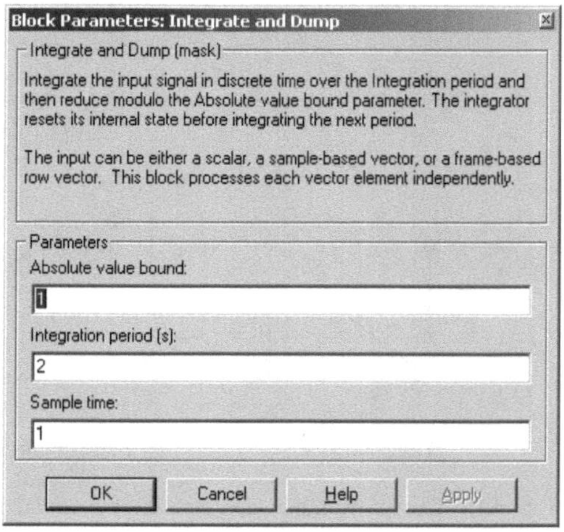

Absolute value bound

This is the modulus by which the integration result is reduced. This parameter must be positive.

Integration period (s)

The first reset time. This is also the time interval between resets.

Sample time

This is the integration sample time. This must not exceed the Integration period.

Examples

Integrate a constant signal whose value is 1 using these parameters:

- Absolute value bound = 5
- Integration period = 7
- Sample time = .5

We can use a Simulink Constant block for the input signal. The Simulink Scope block shows the output below.

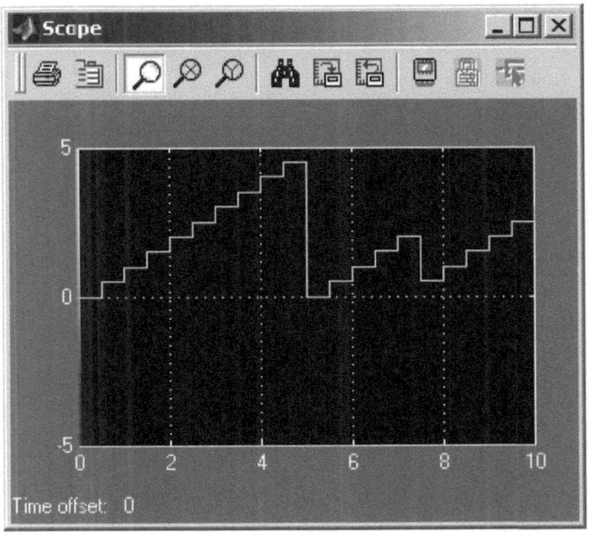

Fig. 3.16 Output of the Integrate and dump filter block for step input as shown in the simulink model editor scope window

References

1. http://www.mathworks.com
2. Singh, R.P., Sapre, S.D., "Communication Ststems – Analog and Digital", TATA McGrawHill, 2005
3. Bhattacharya, Amitabha, "Digital Communication", TATA McGrawHill, 2006

Chapter 4
Digital Baseband Signal Transmitter

4.1 Introduction

This chapter deals with study of transmission of baseband signals being transferred to digital (binary) format and then re-formatting it to transmittable waveform through a digital communication system. We will focus on the mode of formatting of analog signals to digital ones. First we will study the elements of digital baseband communication and their brief functions then we take care of choice of digital formats (i.e. the individual properties and choose which format is appropriate for different types of signaling formats & requirements). We then study different digital formats. The choice of the formats, i.e. method of encoding is really important. In binary signal transmission, the data bits to be transmitted are 1 or 0 and the transmission should be such that, that should be un-erroneously and easily recognizable at the receiver side. In receiver stage, the matched filter reception concept is the optimum detection scheme for baseband transmission of signal through linear, AWGN channels having unlimited bandwidth. But practically, the transmission channels do have finite amount of bandwidth. So, rectangular pulse shape is not suitable for the practical cases. Therefore, transmission of digital signals through band-limited channels needs special attention. Present chapter broadly deals with the line codes [1] for representing digital sequences by pulse waveforms, transmission of rectangular pulses and signaling through band-limited channels.

4.2 Elements of Digital Baseband Communication System

In baseband communication system, the information source can be anything like analog signal, digital signal or textual information. For analog and textual signal the formatting is essential [3]. In case of digital signal, the formatting step may be skipped as shown in the Fig. 4.1. Next a pulse modulation step is there. Before going through the channel another process, channel encoding is to be done for error correction at the receiving section.

A. Das, *Digital Communication*, Signals and Communication Technology,
DOI 10.1007/978-3-642-12743-4_4, © Springer-Verlag Berlin Heidelberg 2010

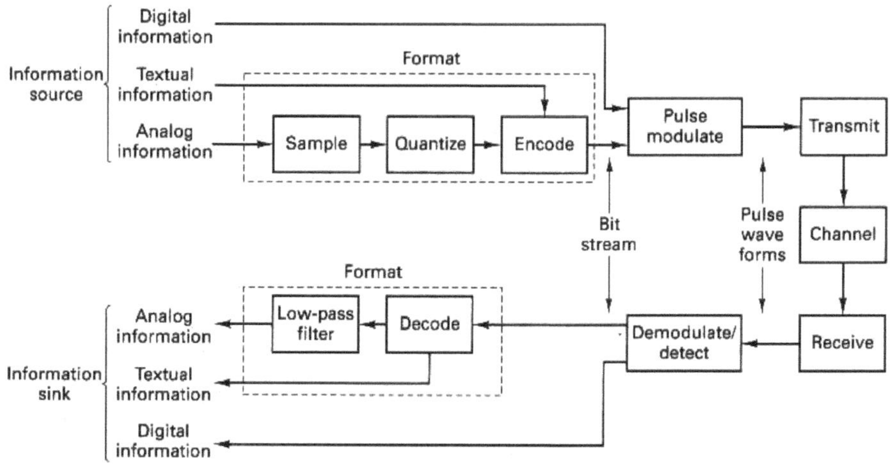

Fig. 4.1 Basic elements of baseband communication system

4.2.1 Formatting

The source output may be either an analog signal, such as an audio or video signal, or a digital signal, such as the output of a teletype machine, that is discrete in time and has a finite number of output characters. In a digital communication system, the messages produced by the source are converted into a sequence of binary digits. The process of efficiently converting the output of either an analog or digital source into a sequence of binary digits is called source encoding or data compression. The sequence of binary digits from the source encoder, which we call the information sequence is passed to the channel encoder. The purpose of the channel encoder is to introduce, in a controlled manner, some redundancy in the binary information sequence that can be used at the receiver to overcome the effects of noise and interference encountered in the transmission of the signal through the channel. The binary sequence at the output of the channel encoder is passed to the digital modulator, which serves as the interface to the communication channel. The same opposite features are available in receiver.

From the discussion of waveform encoding in Chap. 2, we know analog information is formatted using three separate processes.

- Sampling
- Quantization
- Line encoding

4.2.2 Regenerative Repeater

Repeaters are actually amplifiers with suitable frequency response characteristics, placed at regular intervals along the transmission channel to overcome signal attenuation and distortion. Regenerative repeater is one back to back configuration of receiver and transmission in a single package. With lower probability of bit error and large number of regenerative repeaters in use, we can considerably save the transmission power in case of digital signal transmission only (Fig. 4.2).

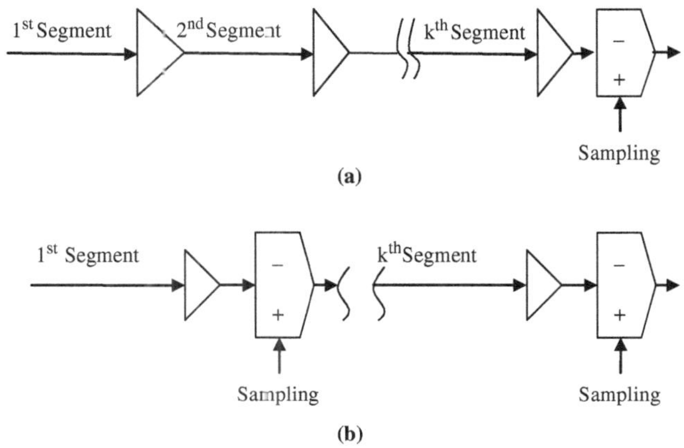

Fig. 4.2 Schematic diagram of a data link with (**a**) non-regenerative repeater, (**b**) regenerative repeater

The repeaters used in a digital link are necessarily regenerative, i.e., they amplify and as well as regenerate the error free transmitted signal from distorted and attenuated received signal through a decision deice at the receiver end. Three basic functions are performed by regenerative repeater as follows [2].

(i) Reshaping of the incoming pulse train using an equalizing filter (to be discussed in detain in Chap. 5).
(ii) Extracting necessary timing information for sampling.
(iii) Decision making based on the state of transmission from sampled values.

As illustrated in above Fig. 4.3, at the first stage of signal processing is *amplification and equalization.* The function of the equalizer is to shape the received pulse so as to compensate the phase and amplitude distortion by the transmission channel. The *timing circuitry* provides a periodic pulse train derived from the received pulse at instants where the SNR is maximized. The *decision making device (regenerator)* which is enabled at the sampling instant determined by the timing circuit makes its decision based on whether or not the amplitude of the quantized signal plus noise exceeds a pre-determined (threshold) voltage level i.e. it makes a decision whether a positive or negative pulse is present. Decision device is actually made of averaging circuit with a comparator with one end connected with the threshold level. In this way the noise and distortion introduced in a repeater space is completely removed provided the distortion is not too long to produce an error in decision making.

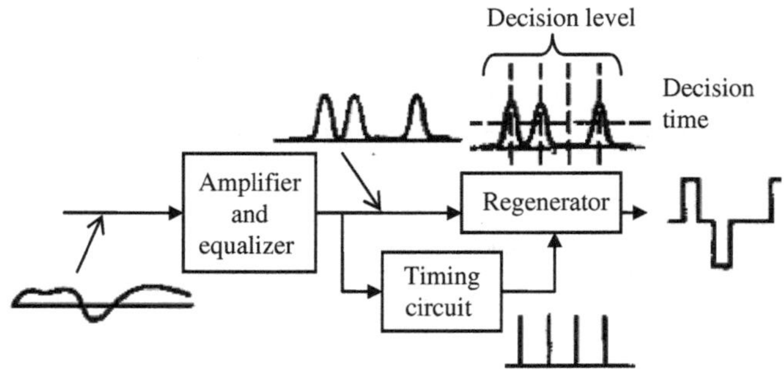

Fig. 4.3 Working principle of regenerative repeater

4.3 Properties and Choice of Digital Formats

The choice of particular form of digital format largely depends on the application of interest. The format of line code must have the following desirable properties.

1. *DC component:* Eliminating the DC energy from the signal's PSD enables the system to be AC coupled. Thus low frequency information could be lost.
2. *Self-clocking:* Bit synchronization or self-clocking is required for any digital communication system. This is different for different formats. There must have some provision of providing the clock signal at the receiving side for faithfully reconstruction of the message signal/ symbol. For bit and frame synchronization, the self clocking is necessary.
3. *Error detection:* Schemes such as duo-binary provides error detection without introducing additional error detection codes. In bipolar signaling, for an example, violation of code rule in detection is considered erroneous.
4. *Bandwidth compression:* Some schemes increase bandwidth utilization by reducing the required bandwidth, thus there is more information per unit bandwidth.
5. *Favorable power spectral density:* This is required to match the spectral occupancy of the signal with the characteristic of the channel. If the channel has poor low frequency response, the signal should not have DC and respective low frequency components.
6. *Differential encoding:* This is an advantageous technique which allows the polarity of the signal to be reversed without affecting the data detection.
7. *Noise immunity:* The various PCM waveforms can be characterized by probability of error vs. SNR. Polar signaling in this count is advantageous over the unipolar signaling.
8. *Transparency:* The word transparency in regarding line coding signifies the capacity of retrieving data regardless of the pattern of 1s and 0s. If data can be reconstructed for every possible data stream representation, the code is known as transparent code. In unipolar transmission, a long string of 0s in the bit stream causes loss of synchronization.

4.4 Line Coding

As discussed, line codes (or transmission code for baseband communication) are the representative techniques of digital sequences by pulse waveforms suitable for baseband transmission. We here consider both binary and M-ary digital baseband transmission technique. By binary we mean one of possible two level output of pulse generator in response to each bit. In the baseband M-ary version of the system, the output of the pulse generator may take on any of the possible M amplitudes with M>2. The blocks of n-message bits are represented by an M-level waveform with $M = 2^n$.

The waveforms broadly can be classified into the following four groups.

1. Non-return to zero (NRZ)
2. Return to zero (RZ)
3. Phase encoded
4. Multilevel binary

The NRZ group is probably the most popular in practical usage. It can again be partitioned into the following sub-groups: NRZ-L (L for level), NRZ-M (M for mark), and NRZ-S (S for space). NRZ-L is used extensively in digital logic circuit. Binary one is represented by one voltage level and binary zero by another. There is a change in level whenever the data change from a one to zero or from zero to one. With NRZ-M, the one, or *mark* is represented by a change in level and a zero or *space* is represented by no change in level. This is sometimes called as *differential encoding.* NRZ-M is used primarily in magnetic tape recording. NRZ-S is the complement of NRZ-M: binary one is represented as no change in level and a zero is represented as change in level.

The RZ waveforms consist of unipolar-RZ, and RZ-AMI. In magnetic tape recording and baseband data transmission we can use these types of codes. With unipolar-RZ, a 'one' is represented by a half-bit-wide pulse and a 'zero' is represented by absence of pulse. With bipolar-RZ, the one and zero is represented by opposite level pulses that are one-half bit-wide as shown in the Fig. 4.4. It means, in each bit interval T_B, there is a pulse present. RZ-AMI (AMI for 'alternate mark inversion') is a signaling scheme used in telephone systems. Here also, the ones are represented by the absence pf pulses.

The phase-encoded group consists of bi-phase-level ($bi - \phi - L$); *and delay modulation* or *miller coding.* The phase encoding schemes are used in magnetic recording systems, optical communications and satellite telemetry systems. In $bi - \phi - L$, a one is represented by a half bit wide pulse positioned during the first half of the bit interval; a zero is represented by a half bit wide pulse positioned during the second half of the bit interval as shown in Fig. 4.4. In $bi - \phi - M$, a transition occurs at the beginning of every bit interval. A one is represented by a second transition one half bit interval later; a zero is represented by no second transition. In case of $bi - \phi - S$, a transition also occurs at the beginning of every bit interval. A one is

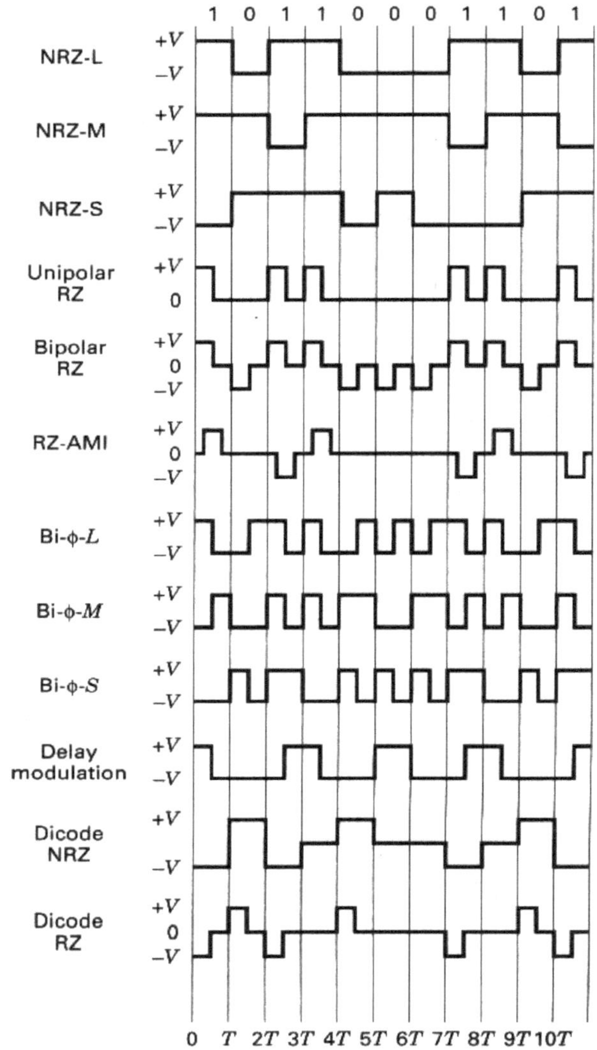

Fig. 4.4 Different line coding formats

represented by no second transition; a zero is represented by a second transition one half bit interval later.

There are many other waveforms which use three voltage/current levels instead of two. Bipolar-RZ and RZ-AMI belong to this group. The group also contains formats called *dicode* and *duobinary*. In dicode-NRZ the one-to-zero or zero-to-one data transition changes the pulse polarity; without a data transition, the zero level is transmitted. On the other side, in dicode-RZ the one-to-zero or zero-to-one data transition generates a half-duration polarity change; otherwise, zero level is transmitted.

4.5 Power Spectrum Density of Different Digital Formats

The flow of studying the power spectral density (PSD) in this section would be 1st studying the PSD for generalized digital signal and then the result will be applied to the typical line codes. Power spectrum can be found by computing the correlation function of the random pulse and then taking the frequency domain (Fourier transform) of it.

The digital bit stream can be considered as random pattern of 1s and 0s. Baseband modulation or digital formatting of line coding signifies only the properly shaping of the pulse representation, i.e., assigning two specific amplitudes for the digit 1 and 0 respectively. Such signal can be written as

$$\begin{aligned} x(t) &= \ldots + a_{-1}p(t - T_B) + a_0\, p(t) + a_1\, p(t + T_B) + \ldots \\ &= \sum_{k=-\infty}^{\infty} a_k\, p(t - kT_B) \end{aligned} \tag{4.1}$$

where a_k represents the amplitude of kth digit, T_B is the bit duration. The pulse shape $p(t)$ should satisfy the condition

$$p(t) = \begin{cases} 1; k = 0, i.e., t = 0 \\ 0; k \neq 0, i.e., t = \pm T_B, \pm 2T_B, \ldots \end{cases} \tag{4.2}$$

To satisfy the condition of Eq. (4.2), $p(t)$ must be a *rect* function located at t=0, i.e., $p(t) = 0$ for $|t| \geq T_B/2$. Equation (4.2) also ensures that, the transmitted digits can be recovered by periodic sampling $x(t)$ at $t = \pm mT_B$, where m is a positive integer. Therefore,

$$x(mT_B) = x(t)|_{t=mT_B} = \sum_k p(mT_B - kT_B) = a_m \tag{4.3}$$

The rate of signaling (r_B) and the bit duration is related as

$$r_B = 1/T_B \tag{4.4}$$

A periodic alternate string of 1s and 0s is now considered as a typical digital bit stream. Here the period $2T_B (= 2/r_B)$ where r_B is the signaling rate. Therefore, the minimum bandwidth required to transmit such a signal is $r_B/2$ Hz, which is the fundamental frequency of the square wave. If the available bandwidth is less than this, the channel cannot be used for successful transmission.

In case of generalized digital signal as described by the Eq. (4.1) is of infinite amount of samples i.e., with infinite amount of energy. Therefore the Fourier transform of the signal does not exist. To solve the problem lets define $x(t)$ in a truncated form as

$$x_T(t) = \begin{cases} x(t); |t| \le T_B/2 \\ 0; \text{otherwise} \end{cases} \tag{4.5}$$

where, T is large but not infinite. Now the Fourier Transform can be defined as

$$X_T(f) = \int_{-\infty}^{\infty} x_T(t) e^{-j\omega t} dt \tag{4.6}$$

If Eq. (4.6) exists, then we can move towards the use of the property of auto-correlation mentioned before. Considering the integral, we get from Eq. (4.6),

$$\int_{-\infty}^{\infty} x_T(t) x_T(t-\tau) dt = \int_{-\infty}^{\infty} |X_T(f)|^2 e^{j\omega\tau} df \tag{4.7}$$

If $T \to \infty$, the integration has no meaning, but if the average is taken and then the limit is applied, we can define $\xi(\tau)$ as

$$\xi(\tau) = \underset{T\to\infty}{Lt} \frac{1}{T} \int_{-\infty}^{\infty} |X_T(f)|^2 e^{j\omega\tau} df \tag{4.8}$$

Hence, we can have the auto-correlation function, interchanging the order of integration as

$$R_x(\tau) = \int_{-\infty}^{\infty} \underset{T\to\infty}{Lt} \frac{E\lfloor |X_T(f)|^2 \rfloor}{T} e^{j\omega\tau} df \tag{4.9}$$

According to W–K theorem, auto-correlation function and PSD ($G_x(f)$) are Furier Transform pair, i.e., $R_x(\tau) = \int_{-\infty}^{\infty} G_x(f) e^{j\omega\tau} df$. Therefore, from Eq. (4.9),

$$G_x(f) = \underset{T\to\infty}{Lt} \frac{E\lfloor |X_T(f)|^2 \rfloor}{T} \tag{4.10}$$

If there are samples expressed with $t = mT_B$, with $m = 0, \pm 1, \pm 2, \ldots, \pm M$, i.e. $(2M+1)$ number of samples, $T = (2M+1)T_B$. Then Eq. (4.10) becomes

$$G_x(f) = \underset{M\to\infty}{Lt} \frac{E\lfloor |X_T(f)|^2 \rfloor}{(2M+1)\,T_B} \tag{4.11}$$

Now,

$$x_T(t) = \sum_{m=-M}^{M} a_m\, p(t - mT_B) \tag{4.12}$$

Taking the F.T. of both sides of Eq. (4.12),

$$X_T(f) = \sum_{m=-M}^{M} a_m P(f) e^{-jm\omega T_B} \tag{4.13}$$

We know,

$$\begin{aligned} |X_T(f)|^2 &= X_T(f).X_T^*(f) \\ &= \sum_{n=-M}^{M} \sum_{m=-M}^{M} a_m\, a_n\, P(f)P^*(f) e^{-jn\omega T_B} e^{jm\omega T_B} \end{aligned} \tag{4.14}$$

Hence,

$$\begin{aligned} G_x(f) &= \underset{M\to\infty}{Lt} \frac{|P(f)|^2}{(2M+1)T_B} \sum_n \sum_m E[a_n\, a_m] e^{-j\omega(n-m)T_B} \\ &= \underset{M\to\infty}{Lt} \frac{|P(f)|^2}{(2M+1)T_B} \sum_n \sum_m R_a(n-m) e^{-j\omega(n-m)T_B} \\ &= \underset{M\to\infty}{Lt} \frac{|P(f)|^2}{(2M+1)T_B} \sum_n \{\Psi(n-K) + \Psi(n-K+1) + \ldots \Psi(n) \ldots \\ &\qquad + \Psi(n+K)\} \\ &\qquad (\text{Taking } \Psi(n-m) = R_a(n-m) e^{-j\omega(n-m)T_B}) \\ &= \underset{M\to\infty}{Lt} \frac{|P(f)|^2}{(2M+1)T_B} \left[\sum_{k=-2K}^{0} (2K+1+k)\Psi(k) + \sum_{k=1}^{2K} (2K+1-k)\Psi(k) \right] \\ &= \underset{M\to\infty}{Lt} \frac{|P(f)|^2}{(2M+1)T_B} \left[\sum_{k=-2K}^{2K} (2K+1-|k|)\Psi(k) \right] \\ &= \underset{M\to\infty}{Lt} \frac{|P(f)|^2}{(2M+1)T_B} \left[(2K+1) \sum_{k=-2K}^{2K} \left(1 - \frac{|k|}{2K+1}\right) R_a(k) e^{-j\omega k T_B} \right] \\ \therefore G_x(f) &= \frac{|P(f)|^2}{T_B} \sum_{k=\infty}^{\infty} R_a(k) e^{-j\omega k T_B} \end{aligned}$$

$$\tag{4.15}$$

Equation (4.15) gives the PSD of a digital signal from its pulse spectrum P(f) and auto-correlation function $R_a(k)$.

Power spectral densities (PSD) of all the popular line codes are now discussed in the preceding subsection.

4.5.1 Unipolar-NRZ

In unipolar-NRZ format, digit 1 is represented by a pulse of amplitude A and digit 0 is represented with a pulse of amplitude 0 (Fig. 4.5). The pulse shape p(t) is the kth signaling element, expressed as along with its F.T.

$$p(t) = \Pi\left(\frac{t - kT_B}{T_B}\right) \text{ and } P(f) = T_B \sin c\,(T_B f)\, e^{-jk\omega T_B} \tag{4.16}$$

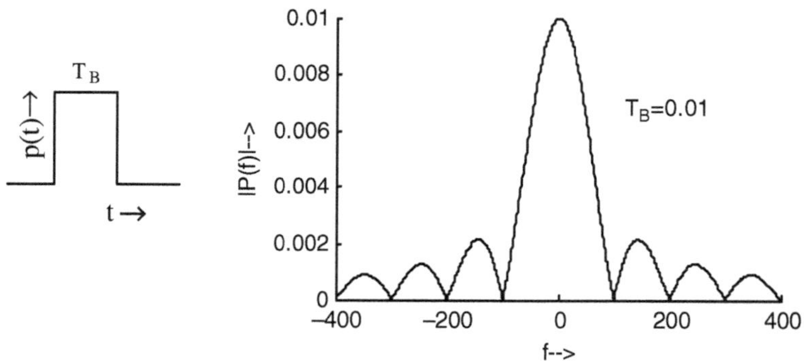

Fig. 4.5 Unipolar-NRZ signaling waveform and its spectra

The amplitude auto-correlation, $R_a(k)$ for such line code is expressed as

$$R_a(k) = E[a_j a_{j-k}]$$

where, $E \underline{\Delta} Ensemble$. Considering equi-probable digits, i.e., $P(a_j = A) = P(a_j = 0) = 0.5$, we can infer

$$\begin{aligned} R_a(0) &= E[a_j a_j] = A^2 \times P(a_j = A) + 0^2 \times P(a_j = 0) = A^2/2 \\ R_a(1) &= E[a_j a_{j-1}] = \{AP(a_j = A) \times AP(a_{j-1} = A)\} + \\ &\quad \{AP\left(a_j = A\right) \times 0.P\left(a_{j-1} = A\right)\} + \\ &\quad \{0.P\left(a_j = A\right) \times A.P\left(a_{j-1} = A\right)\} + \\ &\quad \{0.P\left(a_j = A\right) \times 0.P\left(a_{j-1} = A\right)\} = A^2/4 \end{aligned}$$

Similarly, $R_a(2) = R_a(3) = R_a(4) \ldots\ldots = A^2/2$;
i.e., $R_a(k)|_{k \geq 1} = A^2/2$

Therefore, PSD of this type of line coding is given by

$$G_x(f) = \frac{|T_B \sin c\,(T_B.f)|^2}{T_B}\left\{\frac{A^2}{4} + \frac{A^2}{4}\sum_{k=-\infty}^{\infty} e^{-j\omega k T_B}\right\}$$

Using Poission's sum formula,[1] we can express

$$\begin{aligned} G_x(f) &= \frac{A^2}{4r_B}\sin c^2\left(\frac{f}{r_B}\right)\left\{1 + r_B\sum_{k=-\infty}^{\infty}\delta(f - kr_B)\right\} \\ &= \frac{A^2}{4r_B}\sin c^2\left(\frac{f}{r_B}\right) + \frac{A^2}{4}\delta(f) \end{aligned} \tag{4.17}$$

The Eq. (4.17) signifies, the PSD of unipolar-NRZ is continuous with a discrete impulse at DC (Fig. 4.6).

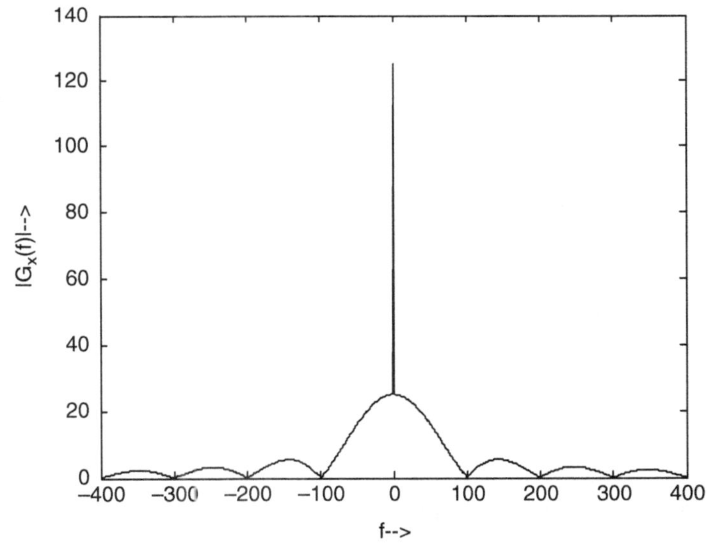

Fig. 4.6 PSD of unipolar-NRZ signaling

4.5.2 Unipolar-RZ

In unipolar-RZ format, the pulse shape p(t) for the kth signaling element and its Fourier Transform is given as

[1]Poission's sum formula states that, $\sum_{k=-\infty}^{\infty} e^{-j2\pi kf/r_B} = r_B \sum_{k=-\infty}^{\infty} \delta\,(f - kr_B)$.

$$p(t) = \Pi\left(\frac{t - kT_B}{T_B/2}\right) \text{ and } P(f) = \frac{T_B}{2} \sin c\left(\frac{T_B}{2} f\right) e^{-jk\omega T_B} \tag{4.18}$$

The Fig. 4.7 shows the pulse shape and the frequency domain representation of the signaling format. Following the procedure described in the above section,

$$R_a(1) = R_a(2) = R_a(3) = R_a(4) \ldots\ldots = A^2/4, R_a(0) = A^2/2$$

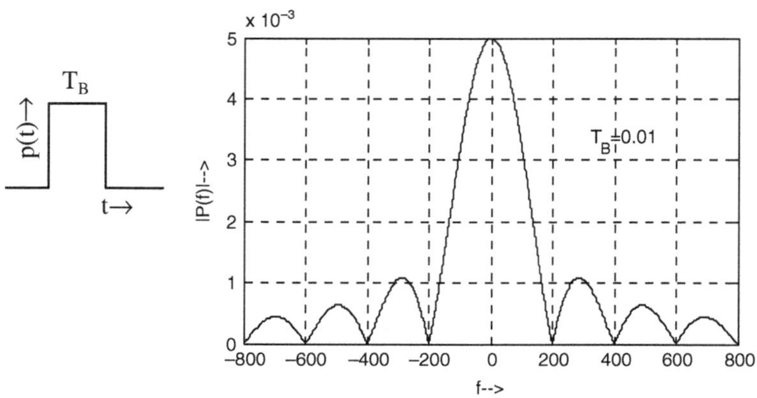

Fig. 4.7 Unipolar-RZ signaling waveform and its spectra

The PSD of the unipolar-RZ signaling is therefore,

$$G_x(f) = \frac{A^2}{16r_B} \sin c^2\left(\frac{f}{2r_B}\right) + \frac{A^2}{16}\delta(f) + \frac{A^2}{16} \sin c^2\left(\frac{n}{2}\right)\delta(f - nr_B); n = 1, 3, 5, \ldots \tag{4.19}$$

From the PSD shown in Fig. 4.8, it's clear that, it can be expressed as a continuous function alongwith a number of discrete impulses.

4.5.3 Polar-NRZ

In polar-NRZ format, the amplitude levels of the pulse shape is always $\pm A$. The pulse shape p(t) and its F.T. is therefore defined as

$$p(t) = \Pi\left(\frac{t - kT_B}{T_B}\right) \text{ and } P(f) = T_B \sin c\,(T_B f)\, e^{-jk\omega T_B} \tag{4.20}$$

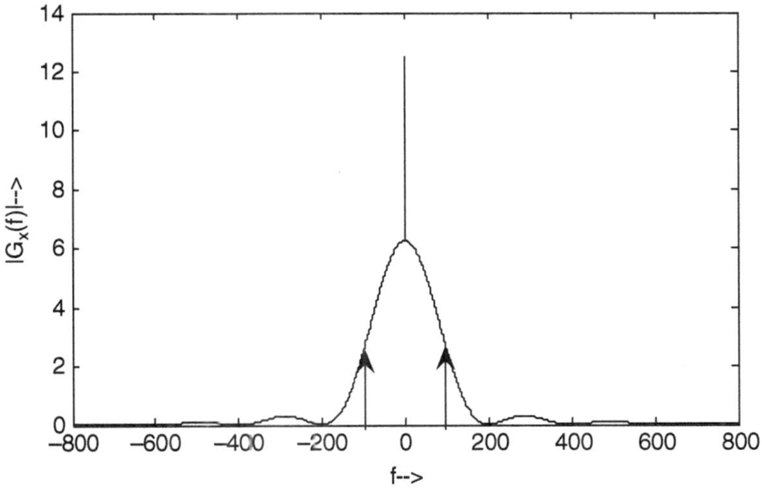

Fig. 4.8 PSD of Unipolar-RZ signaling

Considering equiprobable source digits $R_a(k)$ s are

$$R_a(0) = E\left[a_j a_j\right] = A^2 \times P\left(a_j = A\right) + (-A)^2 \times P\left(a_j = -A\right) = 2 \times A^2/2 = A^2$$

$$\begin{aligned} R_a(1) = E\left[a_j a_{j-1}\right] = &\left\{AP\left(a_j = A\right) \times AP\left(a_{j-1} = A\right)\right\} + \\ &\left\{A.P\left(a_j = A\right) \times -A.P\left(a_{j-1} = -A\right)\right\} + \\ &\left\{-A.P\left(a_j = -A\right) \times A.P\left(a_{j-1} = A\right)\right\} + \\ &\left\{-A.P\left(a_j = -A\right) \times -A.P\left(a_{j-1} = -A\right)\right\} = 0 \end{aligned}$$

$$R_a(k \geq 1) = 0$$

The PSD therefore is given by

$$G_x(f) = \frac{A^2}{r_B} \sin c^2(f/r_B) \tag{4.21}$$

The PSD here is a continuous function of frequency with no discrete component as shown in the Fig. 4.9 below.

4.5.4 Polar-RZ

In polar-RZ format, the pulse shape p(t) and its F.T. is defined as

$$p(t) = \Pi\left(\frac{t - kT_B}{T_B/2}\right) \text{ and } P(f) = \frac{T_B}{2} \sin c\left(\frac{T_B}{2} f\right) e^{-jk\omega T_B} \tag{4.22}$$

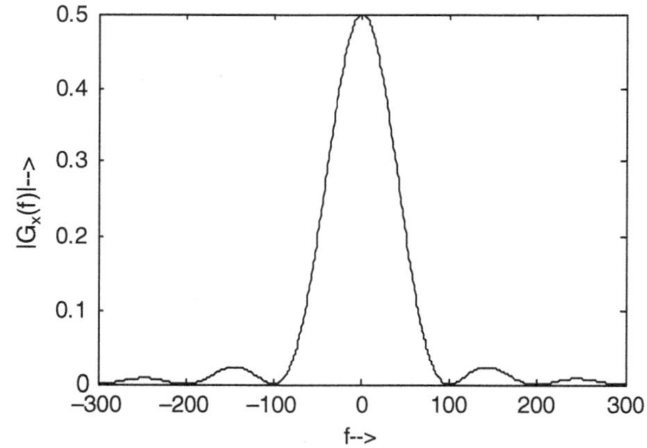

Fig. 4.9 PSD of polar-NRZ signaling

The PSD therefore is given by

$$G_x(f) = \frac{A^2}{4r_B} \sin c^2\,(f/2r_B) \tag{4.23}$$

Here also, the PSD is a continuous function of frequency with no discrete component as shown in the Fig. 4.10 below.

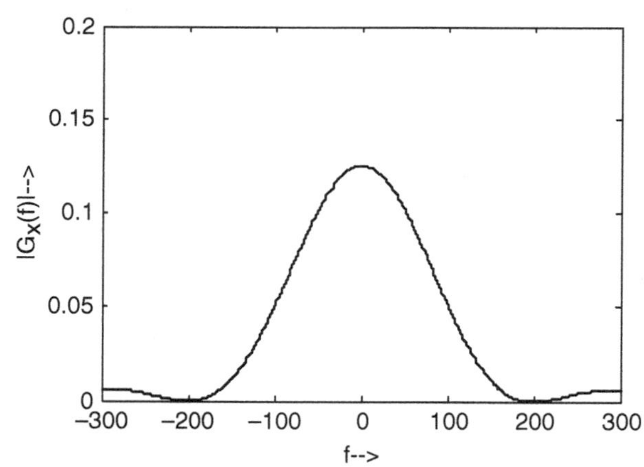

Fig. 4.10 PSD of Polar-RZ signaling

4.5.5 Bipolar-NRZ

In bipolar-NRZ format, the pulse shape p(t) and its F.T. is defined as

$$p(t) = \Pi\left(\frac{t - kT_B}{T_B}\right) \text{ and } P(f) = T_B \sin c\,(T_B f)\, e^{-jk\omega T_B} \tag{4.24}$$

Here, the peculiarity is the probability of the source and the coded digits are not the same. For equi-probable sources, P(0) = P(1) = 1/2. But, as in bipolar signaling, 0 is coded as 0 and 1 si coded as $\pm A$, $P(0) = 1/2$ but, $P(\pm A) = 1/4$. $R_a(k)$ s now can be expressed as

$$\begin{aligned} R_a(0) &= E\left[a_j a_j\right] = A^2 \times P\left(a_j = A\right) + (-A)^2 \times P\left(a_j = -A\right) \\ &\quad + (0)^2 \times P\left(a_j = 0\right) = 2 \times A^2/4 = A^2/2 \\ R_a(1) &= E\left[a_j a_{j-1}\right] = -A^2/4 \\ R_a(k \geq 2) &= 0 \end{aligned}$$

The PSD therefore is given by

$$G_x(f) = \frac{A^2}{r_B} \sin c^2 \left(f/r_B\right) \sin^2\left(\pi f/r_B\right) \tag{4.25}$$

Here also, the PSD is a here two lob and continuous function of frequency with no discrete component (Fig. 4.11).

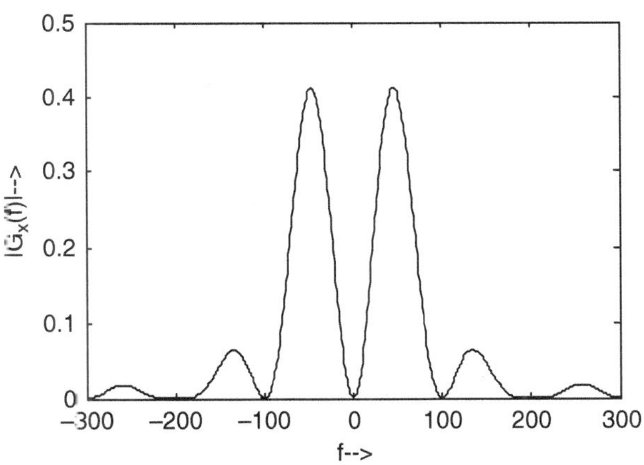

Fig. 4.11 PSD of bipolar-NRZ signaling

PSD of bipolar-RZ can also be obtained easily by this way. Therefore is not mentioned here.

4.5.6 Split-Phase (Manchester)

In this signaling format, p(t) is defined as

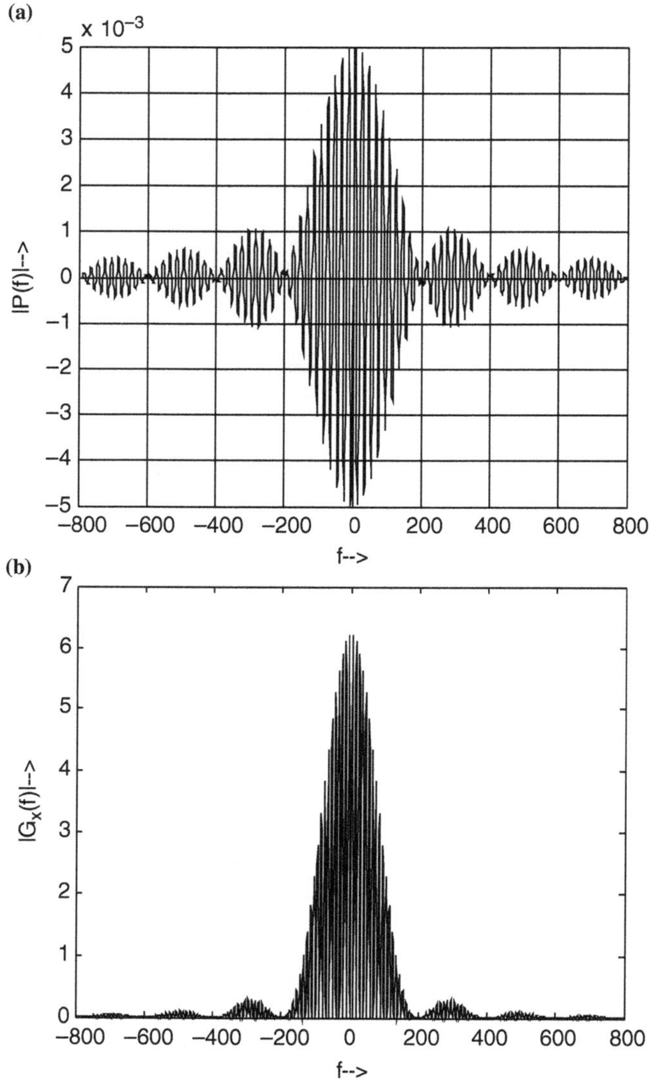

Fig. 4.12 Manchester signaling (**a**) Spectrum of the signaling element, (**b**) PSD

$$p(t) = \pm\left[\Pi\left(\frac{t+\frac{T_B}{4}}{\frac{T_B}{2}}\right) - \Pi\left(\frac{t-\frac{T_B}{4}}{\frac{T_B}{2}}\right)\right] \text{ and the spectrum is}$$

$$P(f) = \pm\left[j2\frac{T_B}{2}\sin c\left(\frac{T_B}{2}f\right)\left(\frac{e^{jk\omega\frac{T_B}{4}} - e^{-jk\omega\frac{T_B}{4}}}{2j}\right)\right] \tag{4.26}$$

In other form,

$$|P(f)| = \frac{1}{r_B} \sin c \frac{f}{2r_B} \sin \frac{\pi f}{2r_B}$$

The $R_a(k)$ s are similar to that of polar signaling, i.e.,

$$R_a(0) = A^2, R_a(1) = R_a(2) = \ldots\ldots 0,$$

Therefore,

$$G_x(f) = \frac{A^2}{r_B} \sin c^2 \frac{f}{2r_B} \sin^2 \frac{\pi f}{2r_B}$$

The waveform and power spectral density (PSD) of Manchester signalling are shown in Fig. 4.12.

References

1. Carlson, A.B., Crilly, P.B., Rutledge, J.C., "Communication Systems – An Introduction to Signals and Noise in Electrical Communication (4th Ed.)", McGraw-Hill International, 2002
2. Chakrabarty, N.B., Datta, A.K., "An Introduction to the Principles of Digital Communications", New Age, 2007
3. Skler, B., "Digital Communications: Fundamentals and Applications (2nd Ed.)", Prentice Hall, 2007

Chapter 5
Equalization

5.1 Inter-Symbol Interference (ISI)

In connection with the digital communication system architecture, inter-symbol interference is an important terminology. As the name suggests, inter-symbol interference (ISI) is the interference of the adjacent bits (in terms of pulses) into the desired pulse shape.

Next, a natural question comes into mind that, why the pulses are interfered by the adjacent pulses and to which extent of neighborhood the effect of interference exists. To answer the question, we are taking the macro level architecture [2, 4] of baseband binary data communication system as shown in Fig. 5.1.

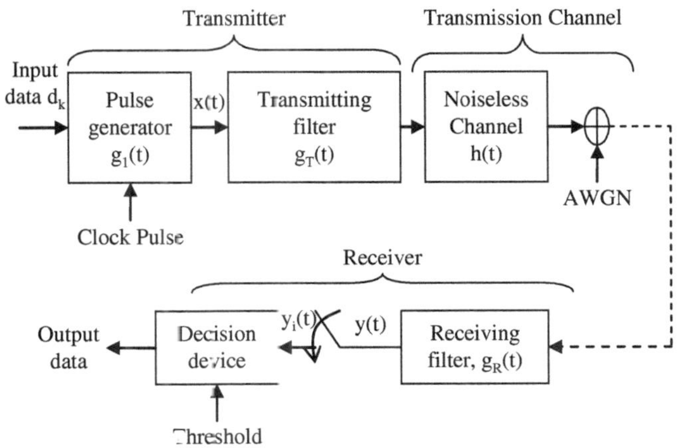

Fig. 5.1 Baseband binary data transmission system

The aforementioned figure is depicting a binary PAM system. The input of the entire system is a binary data sequence $\{d_k\}$ with bit duration T_B. The sequence is applied to a pulse generator to produce a discrete PAM signal as:

A. Das, *Digital Communication*, Signals and Communication Technology,
DOI 10.1007/978-3-642-12743-4_5, © Springer-Verlag Berlin Heidelberg 2010

$$x(t) = \sum_{k=-\infty}^{\infty} a_k s\,(t - kT_B) \tag{5.1}$$

where, *s(t)* denotes the basic pulse, such that normalized *s*(0)=1. The pulse generator block therefore converts the input sequence into polar form as,

$$\begin{aligned} &\text{if} \quad\;\;\; \mathrm{d_k} = 1 \;\; \text{then} \;\; a_k = +1 \\ &\text{and if} \;\; \mathrm{d_k} = 0 \;\; \text{then} \;\; a_k = -1 \end{aligned}$$

As in Fig. 5.1, the shaped polar pulses x(t) are passed through the transmitting filter. Next, the filtered output is transmitted over the transmission channel with impulse response h(t).

The received signal is sampled after proper filtering. At the end, the sampled signal is compared (decision device) with a proper threshold (generally zero) to reconstruct the original binary sequence.

Therefore, the receiving filter output is

$$\begin{aligned} y(t) &= \left\{x(t)^* g_T(t)^* h(t) + n(t)\right\}^* g_R(t) \\ &= x(t)^* \left\{g_T(t)^* h(t)^* g_R(t)\right\} + n(t)^* g_R(t) \\ y(t) &= \mu \sum_{k=-\infty}^{\infty} a_k p\,(t - kT_B) + n'(t) \end{aligned} \tag{5.2}$$

where all the $g_i(t)$ s are the transfer function or impulse response of the respective blocks as shown in Fig. 5.1. μ is the scaling factor of the combined system. p(t) is the convolution of the impulse responses. $n'(t)$ is the convolved noise, i.e., indeed noise.

Now, after sampling, at $t = t_i = iT_B$, and considering the noise free channel,

$$\begin{aligned} y(t_i) &= \mu \sum_{k=-\infty}^{\infty} a_k p\,(t_i - kT_B) \\ &= \mu a_i + \sum_{\substack{k=-\infty \\ k \neq i}}^{\infty} a_k p\,(iT_B - kT_B) \end{aligned} \tag{5.3}$$

The 1st term of Eq. (5.3) signifies the formatted symbol pulse shape of the desired instant/sample. We would be happy, if we get only the 1st term at the receiver terminal. The second term represents the residual effect of all the transmitted symbols, obtained at the time of sampling the ith bit. The effect of neighboring bits would be obviously more than the other bits. This residual effect of other symbols is known as Inter-Symbol Interference.

When a pulse of duration T_B is transmitted through a band limited system, the frequency components contained in the input pulse are differentially attenuated and delayed by the system. Due to this, the pulse appearing at the output is *dispersed* over an interval which is longer than T_B (Fig. 5.2).

Due to this dispersion, the symbol of the ith instant is co-occurred with the delayed and advanced pulses. This will result in Inter-Symbol Interference.

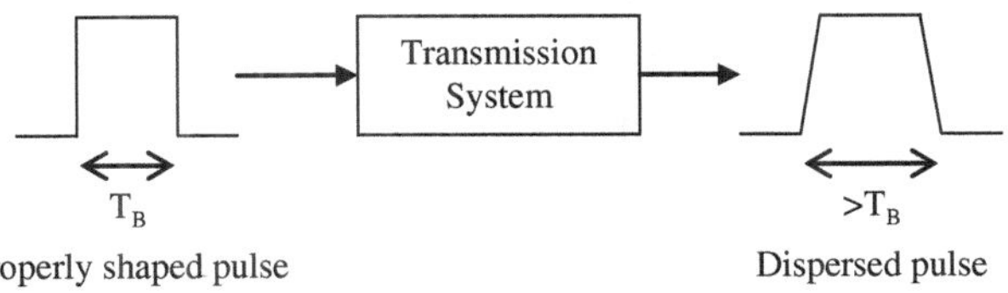

Fig. 5.2 Cause of Inter-Symbol Interference

5.2 Nyquist Criterion for Distortion Less Transmission (Zero ISI)

From the discussion of previous section, it can be inferred that, in absence of ISI, the ith recovered sample at the output should be

$$y(t_i) = \mu a_i$$

To achieve this sample accurately, the transmitting and receiving filters need to be designed properly. The transfer function of the entire transmission system, P(f) needs to be obtained.

The decoding at the receiver end should be such that, the contribution of the weighted pulse i.e., $\sum_{k=-\infty}^{\infty} a_k p\,(iT_B - kT_B)$ at i=k be free from ISI. The said condition can be mathematically modeled as

$$p\,(i-k)\,T_B = \begin{cases} 1 & for \quad i = k \\ 0 & for \quad i \neq k \end{cases} \tag{5.4}$$

where, $p(0) = 1$ due to normalization.

If the received pulse satisfies the Eq. (5.4), then the receiver output given by Eq. (5.3) reduces to

$$y(t_i) = \mu a_i \tag{5.5}$$

This indicates zero ISI in the absence of noise.

5.2.1 Criteria in Frequency Domain

To obtain the transfer function of the entire transmission system, the criterion is to be transformed into frequency domain [3]. A sequence of samples are considered as $p(nT_B)$ where, $n = 0, \pm 1, \pm 2, \ldots$. As depicted in Fig. 5.3, it can be assumed as a sampled signal p(t). The sequence $p(nT_B)$ is obtained by sampling p(t), i.e. by multiplying p(t) by an impulse train $\Delta_{nT_B}(t)$ of period T_B. Therefore, the Fourier transform of $p(nT_B)$ must be convolution of the spectrums of p(t) and $\Delta_{nT_B}(t)$. In Fig. 5.3, the spectrum of $p(nT_B)$ is shown and it's easy to understand that, we'll get a periodic spectrum with periodicity $1/T_B$. The mathematical support is as follows.

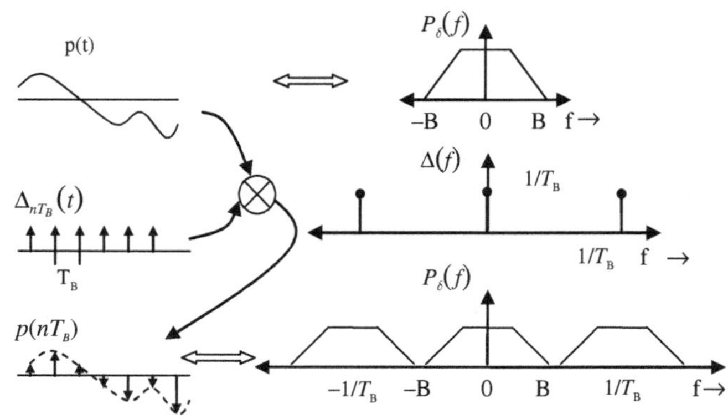

Fig. 5.3 Depiction of Nyquist criteria

The sampled signal is $p(nT_B) = p(t) \times \Delta_{nT_B}(t)$. Let's now expand the Δ_{nTs} (t) signal in Fourier series so that we can study the spectrum of $p(nT_B)$. As Δ_{nTs} (t) is even function of time, by Fourier series [1],

$$\Delta_{nT_B}(t) = a_0 + \sum_{n=1}^{\infty} \left(a_n \cos 2n\pi \frac{1}{T_B} + b_n \sin 2n\pi \frac{1}{T_B} \right)$$

Here,

$$a_0 = \frac{1}{T_B} \int_{T_B} \Delta_{nT_B}(t) dt = \frac{1}{T_B}$$

$$a_n = \frac{2}{T_B} \int_{T_B} \Delta_{nT_B}(t) \cos 2n\pi \frac{1}{T_B} dt = \frac{2}{T_B}$$

$$b_n = 0; \text{ for even function}$$

Therefore,

$$\Delta_{nT_B}(t) = \frac{1}{T_B} (1 + 2\cos\omega_B t + 2\cos 2\omega_B t + 2\cos 3\omega_B t + \ldots..) \tag{5.6}$$

considering $\omega_B = \frac{2\pi}{T_B}$. So, the samples sequence

$$\begin{aligned} p(nT_B) &= p(t) \times \Delta_{nT_B}(t) \\ &= \frac{p(t)}{T_B}\left(1 + 2\cos\omega_B t + 2\cos 2\omega_B t + 2\cos 3\omega_B t + \ldots\right) \end{aligned} \tag{5.7}$$

From the 1st term of the equation above, it is very clear that spectrum of p(nT$_B$) will be similar to P(ω) except the amplitude. Amplitude will be 1/T$_B$ times of that of P(ω). Second term indicates P(ω) shifted in both sides by the amount of ω$_B$ (using negative frequency concept). Similarly 3rd term indicates P(ω) shifted in both sides by the amount of 2ω$_B$ and so on.

Now if we can extract the central spectrum by low pass filter from the assembly of spectrum, then we can easily reconstruct the p(t) signal.

Therefore, the frequency domain representation of $p(nT_B)$ will be $P_\delta(f))$ represented as:

$$P_\delta(f)) = F\left[p(nT_B)\right] = \frac{1}{T_B}\sum_{n=-\infty}^{\infty} P\left(f - n \times \frac{1}{T_B}\right) = R_B \sum_{n=-\infty}^{\infty} P(f - nR_B) \tag{5.8}$$

where, Bit rate$=R_B = 1/T_B$.

Another way to represent the Fourier transform is directly applying the formula of Fourier transform as follows.

$$\begin{aligned} P_\delta(f) &= \int_{-\infty}^{\infty} p(nT_B)\, e^{-j2\pi ft} dt \\ &= \int_{-\infty}^{\infty} p(t) \times \Delta_{nT_B}(t) e^{-j2\pi ft} dt \\ P_\delta(f) &= \int_{-\infty}^{\infty} \sum_{m=-\infty}^{\infty} p(mT_B)\, \delta(t - mT_B) e^{-j2\pi ft} dt \end{aligned} \tag{5.9}$$

Let, $m = i - k$; therefore, if $i = k$; $m = 0$ and if $i \neq k$; $m \neq 0$. Now, putting the condition derived in Eq. (5.4) into Eq. (5.10), for m = 0

$$\begin{aligned} P_\delta(f)) &= \int_{-\infty}^{\infty} p(0)\, \delta(t)\, e^{-j2\pi ft} dt \\ &= p(0) \int_{-\infty}^{\infty} \delta(t)\, e^{-j2\pi ft} dt = p(0) \end{aligned} \tag{5.10}$$

$$\therefore \text{Due to normalization}, P_\delta(f)) = p(0) = 1. \tag{5.11}$$

Now, equating the R.H.S. of Eqs. (5.8) and (5.11),

$$\begin{aligned} \frac{1}{T_B}\sum_{n=-\infty}^{\infty} P(f - nR_B) = 1 \\ Z(f) = \sum_{n=-\infty}^{\infty} P(f - nR_B) = T_B \end{aligned} \tag{5.12}$$

The expression is called as *Nyquist criterion for distortion less transmission.*

5.2.2 Concept of Ideal Nyquist Channel

From the expression of 'Nyquist criterion for distortion less transmission', the L.H.S. of Eq. (5.12) represents a series of shifted spectrums [7]. For n = 0, the L.H.S. corresponds to P(f) and it represents a frequency function with narrowest band which satisfies the aforementioned equation.

Suppose that, the channel has a bandwidth of B Hz. Then, $P(f) = 0$, for $|f| > B$. We distinguish three cases as follows.

1. In this case, $T_B < 1/2B$ or equivalently, $1/T_B > 2B$. Since, $\sum_{n=-\infty}^{\infty} P(f - nR_B)$ contains non-overlapping replicas of $P(f))$ separated by $1/T_B$ as in Fig. 5.3, there is no choice of P(f) to ensure the condition of Eq. (5.12) in this case. Thus, we cannot design a zero ISI system considering case 1.
2. In this case, $T_B = 1/2B$ or equivalently, $1/T_B = 2B$(Nyquist rate). Here, the replicas of $P(f))$ are about to overlap, as they are separated by $1/T_B$ as in Fig. 5.4. Therefore, there exists only one $P(f))$ that results in $P_\delta(f)) = 1$, i.e., $\sum_{n=-\infty}^{\infty} P(f - nR_B) = T_B$as

$$P(f)) = \begin{cases} T_B; & |f| < B \\ 0; & \text{otherwise} \end{cases} \tag{5.13}$$

In other words,

$$P(f)) = T_B \Pi\left(\frac{f}{2B}\right) = \frac{1}{2B} rect\left(\frac{f}{2B}\right),$$

which results in

$$p(t) = F^{-1}\left[P(f))\right] = \sin c\left(\frac{t}{T_B}\right) = \sin c\,(2Bt) \tag{5.14}$$

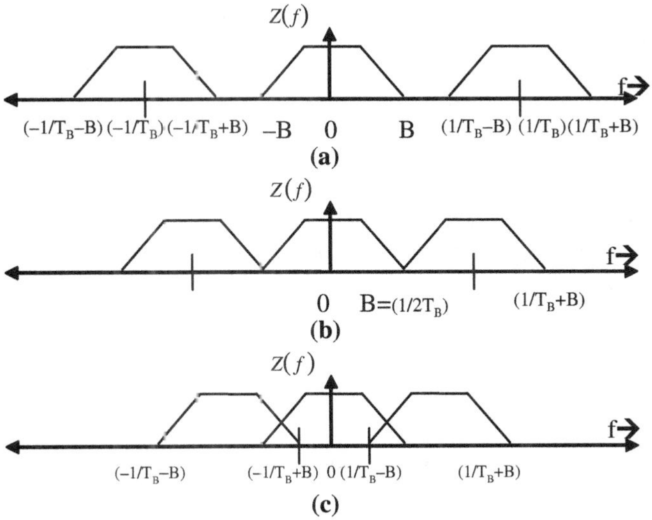

Fig. 5.4 Plot of Z(f) for three cases: (**a**) $T_B < 1/2B$ (**b**) $T_B = 1/2B$, (**c**) $T_B > 1/2B$

This signifies that, the smallest value of T_B for which transmission with zero ISI is possible is $T_B=1/2B$; for this p(t) must be a sinc function. The difficulty is, p(t) is con-causal and hence non-realizable practically.

3. In this case, $T_B > 1/2B$ or equivalently, $1/T_B < 2B$. Here, the replicas of $P(f)$) are totally overlapped. In this case, there exists an infinite number of choices of $P(f)$), such that $Z(f)) \equiv T_B$.

5.2.3 Limitations of Ideal Solution: Raised Cosine Spectrum

The concept of ideal channel says, for satisfying the Nyquist criterion for distortion less transmission (Eq. 5.12), P(f) must be a *rect* function, i.e., a sharp cur of is required at f = B. But, by practical filter, the sharp cut off is not realizable. The only solution is transforming spectrum P(f) to a smoothed version such that, the Nyquist criterion for distortion less transmission is maintained [6, 7].

To handle the situation, particular pulse spectrum that has been widely used in practice is raised cosine spectrum. The raised cosine frequency characteristic is given as

$$P_{rc}(f) = \begin{cases} T_B; & 0 \le f \le (1-\alpha)/2T_B \\ \frac{T}{2}\left[1+\cos\left\{\frac{\pi T_B}{\alpha}\left(|f|-\frac{1-\alpha}{2T_B}\right)\right\}\right]; & \frac{1-\alpha}{2T_B} \le |f| \le \frac{1+\alpha}{2T_B} \\ 0; & |f| > \frac{1+\alpha}{2T_B} \end{cases} \quad (5.15)$$

where, α is the *roll-of factor* of range $0 \le \alpha \le 1$. The bandwidth occupied by the signal beyond the Nyquist frequency is called as *excess bandwidth*, expressed as a percentage of Nyquist frequency. For example, when $\alpha = 0.5$, the excess bandwidth is 50%; when $\alpha = 1$, the excess bandwidth is 100%. The pulse p(t) having the raised cosine spectrum is

$$p(t) = \sin c\left(\frac{t}{T_B}\right)\frac{\cos\left(\pi\alpha t/T_B\right)}{1 - 4\alpha^2 t^2/T_B^2} \tag{5.16}$$

Figure 5.5 illustrates the raised cosine spectral characteristics and the corresponding pulse shapes for $\alpha = 0, 0.5, 1$. It's clear from the Fig. 5.5b, that, for $\alpha = 0$, the pulse reduces to $p(t) = \sin c\,(t/T_B)$ and the symbol rate is $R_B = 1/T_B = 2B$. When $\alpha = 1$, the symbol rate is $R_B = 1/T_B = B$. In general, the tails of $p(t)$decay as $1/t^3$ for $\alpha > 0$.

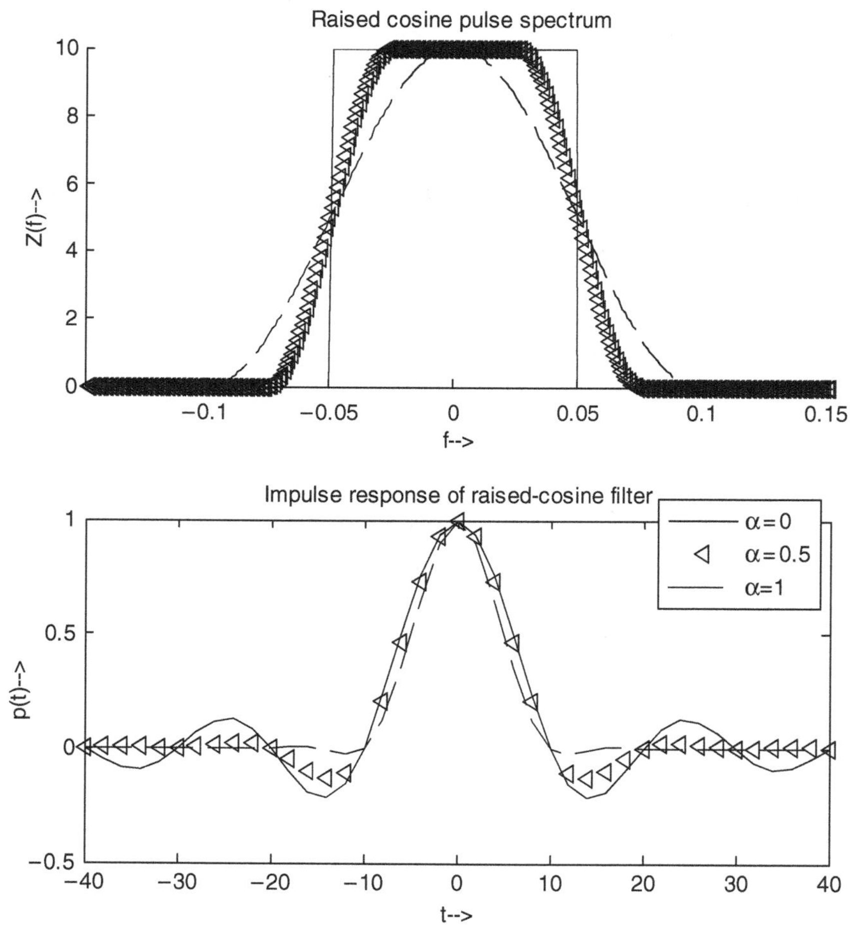

Fig. 5.5 (**a**) Transfer function/spectrum, (**b**) Impulse response of raised cosine pulse filter

Due to the smooth characteristics of raised cosine transform, it is possible to design practical filters for both the transmitter and the receiver the approximate the desired frequency response.

5.3 Eye Pattern

To analyze ISI, eye pattern is a very useful concept of experimental display. In cathode ray oscilloscope (C.R.O.), the eye pattern is displayed [2, 3, 4].

Due to the dispersive nature of the transmission system alongwith the communication channel, inter-symbol interference occurs. ISI distorts the pulse shapes of the transmitted symbols both in terms of attenuation and severely in terms of delay.

Therefore, the transition of each symbol cannot be ensured at each termination end of bit duration. So, the received signal becomes distorted as shown in the Fig. 5.6a. To measure the amount of ISI and its effect on different factors of signal representation, study of eye pattern is important.

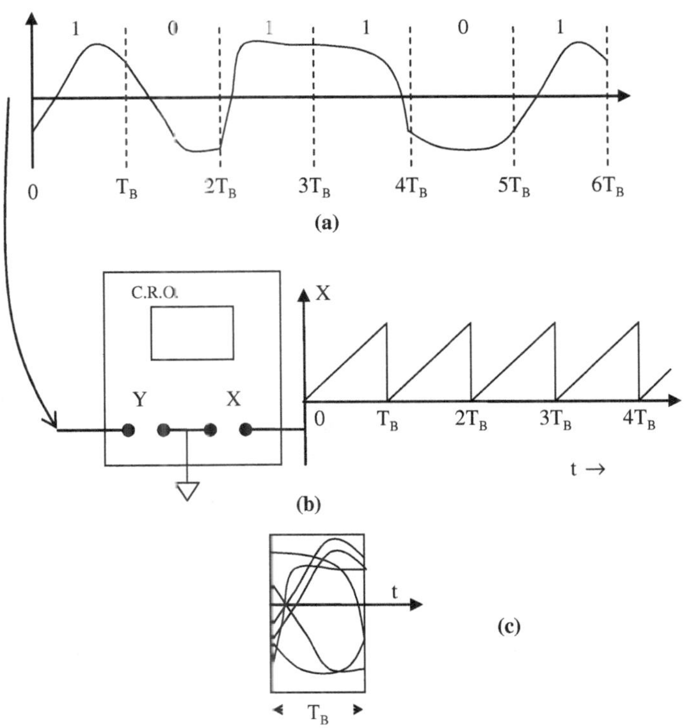

Fig. 5.6 (**a**) Distorted binary wave, (**b**) CRO connections, (**c**) Eye pattern on overlap of distorted binary wave window (as seen in CRO screen)

Each window of received binary wave of width T_B is overlapped visually on the screen of the CRO. Here, time axis is transformed into a new pseudo time (dependent on real time) axis is obtained and the received signal is plotted against the transformed function. Time is here using identity transformation from 0 to T_B. Next, as 't' increases, transformed X value would repeat the value 0 to T_B. By this way, the aforementioned windows are overlapped. The overlapped pattern forms a shape of a human eye (Fig. 5.6c). So, the pattern formed by this way is termed as the *Eye Pattern* (Fig. 5.7).

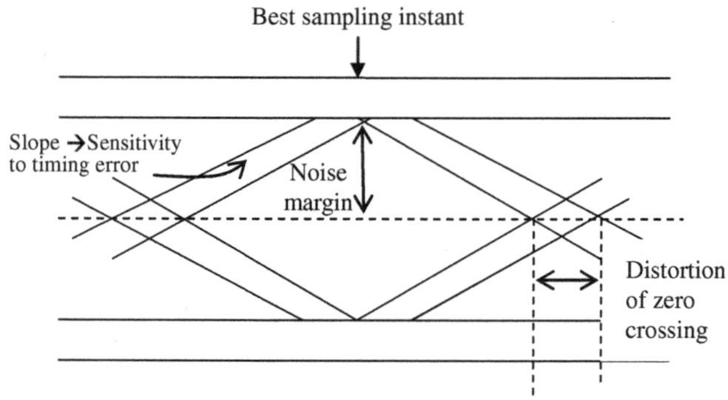

Fig. 5.7 Interpretation of eye pattern

5.3.1 Information Obtained from Eye Pattern

1. The width of the eye opening: *It defines the time interval over which the received wave can be sampled, without an error due to ISI.*
2. The middle of the eye lid with maximum eye opening: *It is the best sampling instant.*
3. The rate of closure of the eye: *It defines the sensitivity of the system to the timing error.*
4. The eye is completely close: *It signifies severe effect of ISI and it is impossible to overcome the combined effect of ISI and noise.*
5. Squinted (asymmetric) eye: *It signifies non-linear transmission distortion.*

5.4 System Design for Known Channel

To design the transmitting and receiving filter characteristics, we 1st assumed a case where the channel frequency response characteristics H(f) is known. From the Fig. 5.1, getting the transfer functions of the respective systems,

$$G_T(f)\,H(f)\,G_R(f) = P_{rc}(f)\,e^{-j2\pi f t_0};\ |f| \le B \tag{5.17}$$

where, t_0 is the time delay to ensure the realizability of the transmitting and receiving filter. To satisfy the Eq. (5.17), suppose

$$G_T(f) = \frac{\sqrt{P_{rc}(f)}}{H(f)} e^{-j2\pi f t_T}; |f| \le B \quad (5.18)$$

$$G_R(f) = \sqrt{P_{rc}(f)} e^{-j2\pi f t_R}; |f| \le B \quad (5.19)$$

where, the delay $t_0 = t_T + t_R$, t_T and t_R are the delays at the transmitting and the receiving side.

Now, the transmission of digital PAM system is in consideration. The aim is to design transmitting and receiving filter for zero ISI at the sampling instant [5, 6]. In practice, the frequency response characteristics of the channel is either unknown or the response is time-variant. For those cases, the frequency response of receiving filter $G_T(f)$ is matched with $G_R(f)$ such that

$$|G_T(f)|\, |G_R(f)| = P_{rc}(f) \quad (5.20)$$

From Eq. (5.3), the output of the receiving filter after sampling is

$$y(t_i) = \mu a_i + \sum_{\substack{k=-\infty \\ k \ne i}}^{\infty} a_k p(iT_B - kT_B)$$

The second term represents ISI. In practical system, it is assumed reasonably that ISI affects a finite number of symbols neighbor to the desired one. Hence, if it is assumed that $a_k = 0$ for $k < -L_1$ and $k > L_2$ where L_1 and L_2 are finite positive integers. Therefore, ISI observed at the output may be viewed as passing the data sequence through a tapped delay type filter as shown in Fig. 5.8. This filter is known as *equivalent discrete time channel filter.*

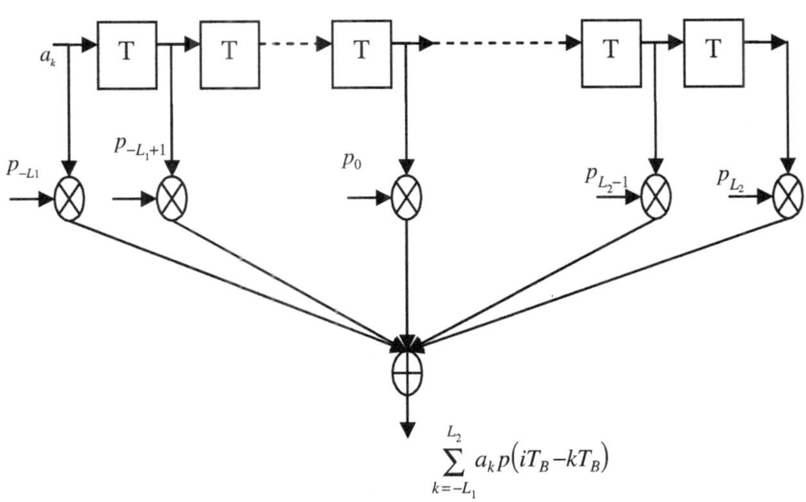

Fig. 5.8 Equivalent discrete time channel filter

5.5 Linear Equalizer

Definition of Equalizer: An equalizer is a filter that compensates for the dispersion effect of the transmission channel.

From the definition of equalizer, we can infer that equalizer is the inverse filter with respect to the channel. If we express the entire communication system shown in Fig. 5.1 in frequency domain, to compensate the effect of channel dispersion, a new filter (equalizer) is to be appended to maintain zero ISI, as shown in Fig. 5.9. Since, the frequency response of transmission and receiving filter are matched to raised cosine pulse spectrum, the frequency response of the equalizer will be only the inverse of the channel response characteristics as

$$G_E(f) = \frac{1}{H(f)} = \frac{1}{|H(f)|\, e^{j\theta_H(f)}} = \frac{1}{|H(f)|} e^{-j\theta_H(f)};\ |f| \leq B \tag{5.21}$$

where,

$$\theta_E(f) = \text{equalizer phase characteristic,}$$
$$\theta_H(f) = \text{channel phase characteristic}$$

and $\theta_E(f) = -\theta_H(f)$. As this inverse channel filter forces the ISI to be zero at the sampling time $t_i = iT_B$, the equalizer is called as *zero forcing equalizer.*

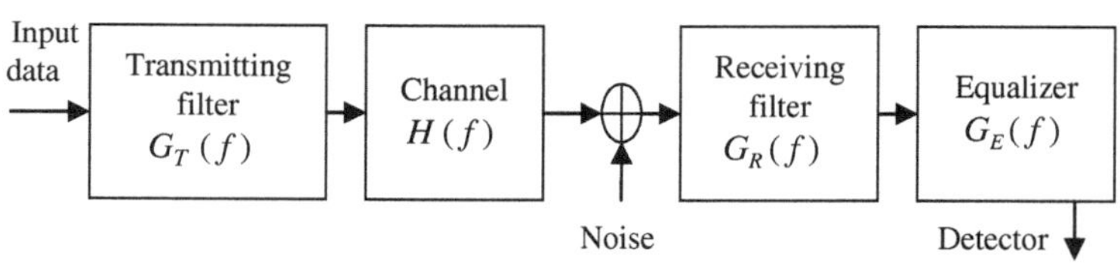

Fig. 5.9 System representation with equalizer

5.5.1 Linear Transversal Filter

Form the discussion of previous section, its clear that, in practical channels the ISI is limited to finite number of neighboring samples. Therefore, practically FIR (Finite Impulse Response) filters or linear traversal filters with adjustable tap gains $\{c_n\}$ may be used to realize the linear equalizer. The maximum time delay τ is equal to the symbol interval.[1] The impulse response of the FIR equalizer is

[1] When the time delay τ is exactly equal to the symbol interval T_B, the FIR equalizer is called as *symbol spaced equalizer.*

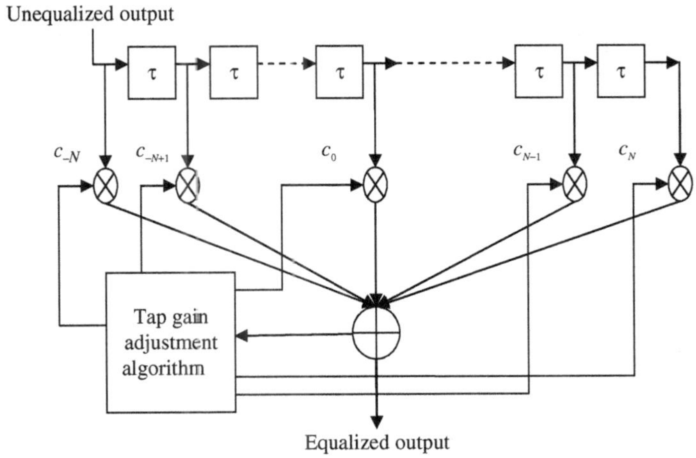

Fig. 5.10 Linear transversal filter

$$g_E(t) = \sum_{n=-N}^{N} c_n \delta(t - n\tau) \tag{5.22}$$

and the corresponding frequency response is

$$G_E(f) = \sum_{n=-N}^{N} c_n e^{j2\pi f n\tau} \tag{5.23}$$

Since, $P(f) = G_T(f)H(f)G_R(f)$, and p(t) is the inverse Fourier transform of P(f), the equalized output signal pulse

$$q(t) = \sum_{n=-N}^{N} c_n p(t - n\tau) \tag{5.24}$$

Applying the zero forcing condition at $t=mT_B$,

$$q(mT_B) = \sum_{n=-N}^{N} c_n p(mT_B - n\tau); m = 0, \pm 1, \pm 2, \ldots \pm N \tag{5.25}$$

Since, there are 2N+1 equalizing coefficient, we can control only 2N+1 sampled values of q(t). Specially,

$$q(mT_B) = \sum_{n=-N}^{N} c_n p(mT_B - n\tau) = \begin{cases} 1, m = 0 \\ 0, m = \pm 1, \pm 2, \ldots \pm N \end{cases} \tag{5.26}$$

Thus, set of (2 N + 1) linear equations are obtained for coefficients of *zero forcing equalization*. Equation (5.26) also may be expressed in matrix form as

$$Pc = q \tag{5.27}$$

where, P is a $(2N + 1) \times (2N + 1)$ matrix with elements $\{p\,(mT_B - n\tau)\}$, c is the (2 N+1) coefficient vector, and q is the (2 N+1) column vector with non-zero elements.

5.6 Adaptive Equalizer

From the previous sections, we have found that, by adjusting the tap gains, proper equalization can be obtained. For practical systems, the situation is somewhat more critical. In practice, the channel response is time-variant and may be non-linear. Therefore, the tap gains need to be adjusted automatically time to time. These type equalizers with the provision of automatic and adaptive tap-gain adjustment are the *adaptive equalizers*. As shown in the Fig. 5.11, 16 point QAM signal is transmitted through a dispersed channel. The eye diagram suggests, proper choice of sampling instant is almost impossible in this case, and in the constellation diagram also, it is showing more than 16 points in the I-Q plane. After passing through an adaptive

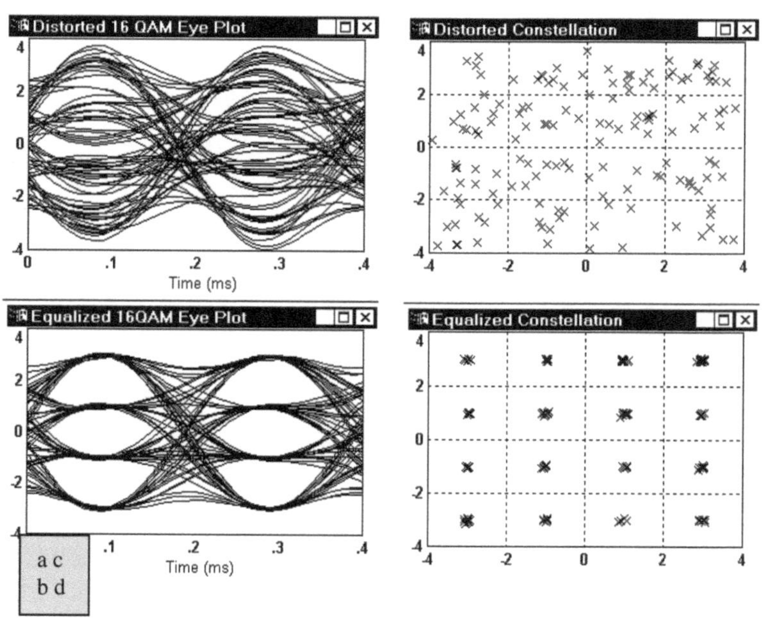

Fig. 5.11 (**a**) and (**c**) Eye plot and constellation diagram of 16 point QAM passed through distorted channel, (**b**) and (**d**) Eye plot and constellation diagram of the signal when passed through an adaptive equalizer

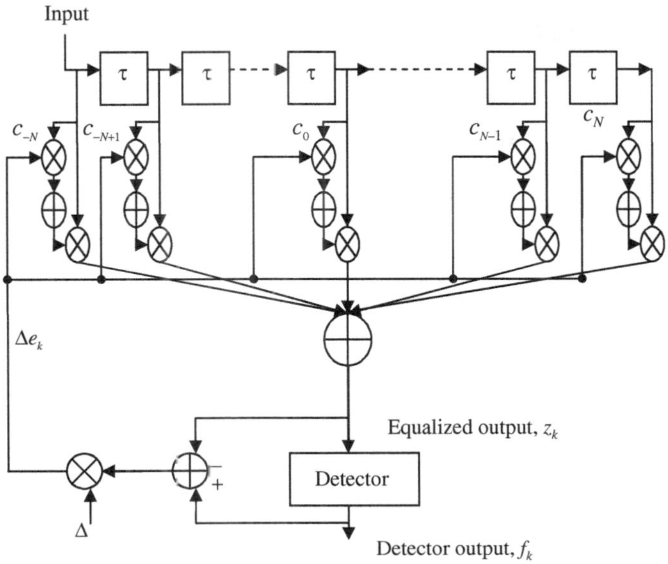

Fig. 5.12 Linear adaptive equalizer

equalizer, both the eye diagram and the constellation diagram improve. Hence, the signal detection, extraction and reconstruction become successful [7].

There are two modes of operations in adaptive equalizer, namely *training mode* and *decision-directed mode.* Firstly, the system is trained by the transmission of a known pseudo-random sequence $\{f_m\}$ over the channel. At the detector side, the filter uses the known sequence to adjust its coefficients. After the initial adjustment, the equalizer changes over to the decision-direct mode, in which case the decisions at the output of the detector are sufficiently reliable so that, the error signal is formed by computing the difference between the detector output and the equalizer output, i.e.,

$$e_k = \tilde{f}_k - z_k \tag{5.28}$$

Here, $\tilde{f}_k$ is the output of the detector. The error is then scaled by a step size Δ, to adjust the tap-gain or coefficient of the equalizer. The scaled error Δe_k multiplies the received signal $\{y\,(kT_B - n\tau)\}$ at the (2 N+1) taps. The products $\Delta e_k y\,(kT_B - n\tau)$ at the (2 N+1) taps are then added to the previous values of tap gain to update the coefficients, as shown in the Fig. 5.12.

A rule of thumb for selecting the step size parameter to ensure convergence of the equalizing system for slow varying (time varying) channels is

$$\Delta = \frac{1}{5\,(2N+1)\,P_R} \tag{5.29}$$

where P_R is the received signal plus noise power, (2 N+1) is the total number of delay taps or no of coefficients.

References

1. Das, A., "Line Communication System – Telecommunication Switching Approach (2nd Ed.)", New Age International Publishers, 2010
2. Chakrabarti, P., "Principles of Digital Communication", Dhanpat Rai and Co, 2008
3. Bedrosian, E., "Spectral Conservation by Efficient Channel Utilization", *Rand Corporation Report*, WN-9255-ARPA, Contract DAHC-15-73-C-0181. Santa Monica, California, October 1975
4. Sharma, S., "Communication Systems – Analog and Digital", S. K. Kataria and Sons, 2009
5. Korn, I., "Principles of Coherent Communications", McGraw-Hill Book Co, 1966
6. Skler, B., "Digital Communications: Fundamentals and Applications (2nd Ed.)", Pearson Education, 2001
7. Proakis, J.G., Salehi, M., "Fundamentals of Communication Systems", Pearson Education, 2006

Chapter 6
Digital Modulation Techniques

6.1 Introduction

Telecommunication suffers from a plenty of problems both in the transmitter and in the receiver end. To maintain realizable antenna height, to provide the facility of multiplexing and channel separation and to get higher level of noise immunity, a number of modulation schemes are employed both in the case of analog and digital communication. Digital modulation signifies modification of three basic properties of analog carrier signal according to the analog corresponding representation of the digital data. A single tone analog carrier signal can be expressed as:

$$s(t) = A \sin(\Omega t + \Phi) \tag{6.1}$$

where,

A = Amplitude of the carrier signal
Ω = Frequency of the carrier signal, and
Φ = Initial Phase of the carrier signal

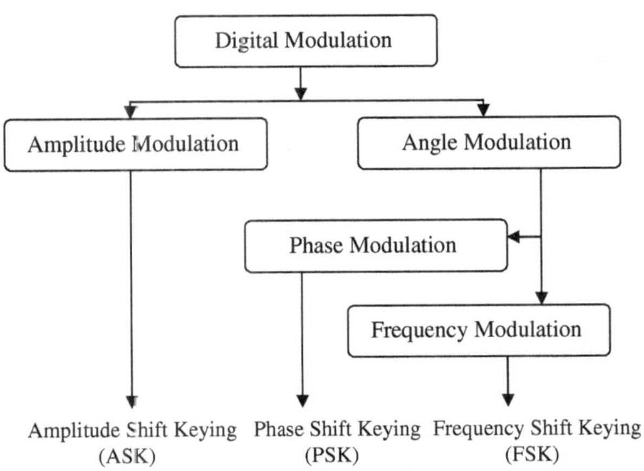

Fig. 6.1 Classification of digital modulation schemes

A. Das, *Digital Communication*, Signals and Communication Technology,
DOI 10.1007/978-3-642-12743-4_6, © Springer-Verlag Berlin Heidelberg 2010

These three are the basic parameters of any signal. Therefore we can modify/modulate any one of these three parameters with respect to the modulating signal or the digital bit stream. Combining the frequency and phase part, we are also getting a time dependent angle part $\theta(t) = (\Omega t + \Phi)$. So, angle modulation is said to be the mother modulation scheme of frequency and phase modulation. The classification is as shown in Fig. 6.1.

6.2 Amplitude Shift Keying (ASK)

Amplitude Shift Keying is the simplest form of digital modulation. We can start understanding ASK referring the prior knowledge of AM [6] in analog modulation domain. In AM (Amplitude Modulation), the amplitude of carrier signal is modulated with respect to the instantaneous amplitude of the baseband signal. There, we get an envelope over the modifying amplitude of the carrier signal as shown in the Fig. 6.2. As we have used sinusoid baseband signal, we are getting sinusoid envelope (shape of change of amplitude of the modulated carrier). Here, the maximum and minimum amplitude of the carrier becomes A_c+mA_c and A_c-mA_c respectively, where A_c is the carrier amplitude and m is the modulation index defined by the ratio of amplitude of the baseband to the carrier.

Now, we can use square type baseband signal using the same carrier, for amplitude modulation. The result would be same, i.e., the envelope will be square type (Fig. 6.3).

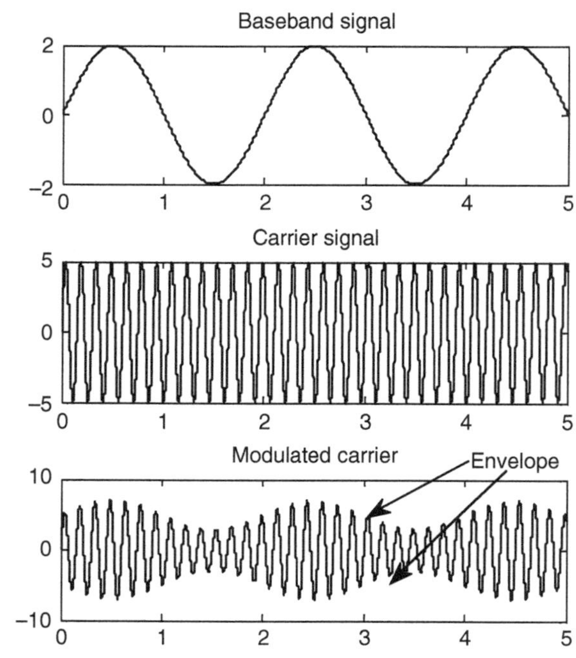

Fig. 6.2 Amplitude modulation (analog modulation) with sinusoid baseband

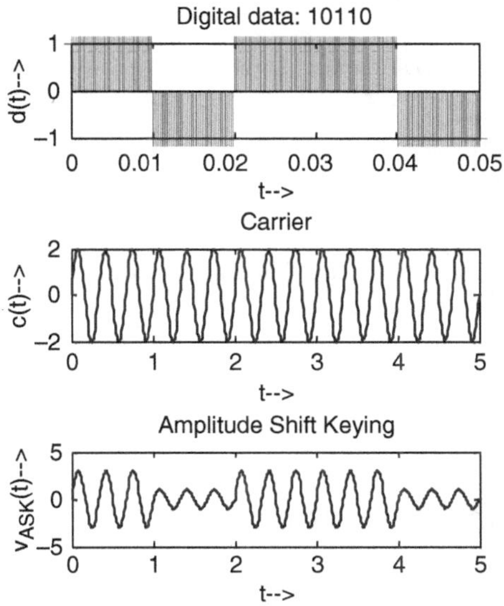

Fig. 6.3 ASK (Digital) as correspondence of AM (Analog)

To make a correspondence with AM, digital baseband signal is represented in bipolar format denoted as d(t), where

$$\begin{aligned} \mathrm{d(t)} \quad &= +1\,\mathrm{V},\ \text{for logic 1} \\ &= -1\,\mathrm{V},\ \text{for logic 0} \end{aligned} \tag{6.2}$$

Here also, high frequency sinusoid is being used as carrier.

The resultant waveform is shown in Fig. 6.3.

Here, we can see from the above figure

$$\begin{aligned} \mathrm{V_{max}} &= \mathrm{A_c(1+m)},\ \text{and} \\ \mathrm{V_{min}} &= \mathrm{A_c(1-m)} \end{aligned} \tag{6.3}$$

where m is the modulation index. No need to mention that, the value of $\mathrm{V_{max}}$ and $\mathrm{V_{min}}$ are identical to that of AM in analog domain.

6.2.1 Mathematical Model

According to the above discussion, the scheme of ASK modulation can therefore be described in a tabular form as shown in Table 6.1.

The last column can be rewritten as,

$$\varphi_{ASK}(t) = A_c\{1 + m(+1)\}\sin\omega_c t \quad if \quad d(t) = +1 \tag{6.4}$$

$$\varphi_{ASK}(t) = A_c\{1 + m(-1)\}\sin\omega_c t \quad if \quad d(t) = -1 \tag{6.5}$$

Table 6.1 ASK waveform definition

Data to be transmitted [d(t)]	Analog equivalent definition	ASK
Binary 1	+1 V	$A_c(1+m)\sin\omega_c t$
Binary 0	−1 V	$A_c(1-m)\sin\omega_c t$

Combining Eqs. (6.4) and (6.5), a generalized expression can be found as

$$\varphi_{ASK}(t) = A\left\{1 + md(t)\right\}\sin\omega_c t \tag{6.6}$$

Mathematically, the scheme can therefore be modeled as the Eq. (6.6).

6.2.1.1 On–Off Keying (OOK)

By ASK the two amplitudes of carrier signals are the representatives of two binary digits. To closely notice, binary digit zero is transmitted in terms of one sinusoid of lower amplitude. But, it is increasing the antenna power by no means. If even no signal is transmitted to represent zero bit, then also the envelope would be identical to that of shown in the Fig. 6.3. Therefore, 100% modulation (m=1) is employed to get more efficient transmission. Hence, carrier signal is switched on and off by the digital bit stream (d(t)). This scheme is called as on–off keying (OOK). Referring Eqs. (6.4) and (6.5), in the modified scheme of ASK modulation, i.e., putting m=1,

$$\varphi_{OOK}(t) = A_c\left\{1 + 1\,(+1)\right\}\sin\omega_c t = 2A_c\sin\omega_c t \quad if \quad d(t) = +1 \tag{6.7}$$

$$\varphi_{OOK}(t) = A_c\left\{1 + 1\,(-1)\right\}\sin\omega_c t = 0 \quad if \quad d(t) = -1 \tag{6.8}$$

The OOK waveform is shown in the Fig. 6.4. Though, OOK is a special type of ASK, in practical cases OOK is used to get ASK waveforms. Therefore, from this time onwards, ASK would be signified as OOK.

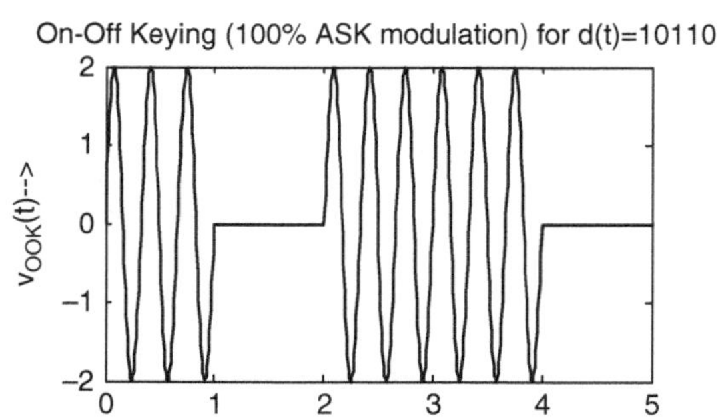

Fig. 6.4 On–Off keying waveform

6.2.2 ASK Modulator

As discussed in the previous section, the digital information is the controller of the amplitude of the fixed frequency sinusoid. When logic 1 is to be transmitted, output of the modulator is sine wave with amplitude $A_c(1+m)$, similarly logic 0 is transmitted in terms of a sinusoid (i.e. single tone or monotone signal) of amplitude $A_c(1-m)$. So, there exist two different sine waves, one whose amplitude is more than the other one. Thus, the above scheme can be realized using ASK modulator as presented in the Fig. 6.5.

Here, when d(t) is logic 1, the switch moves to position 1 and when d(t) is logic 0, it moves over to position 0. Thus, the process of modulation is simply a scheme of switching. Now, the switch can be realized using any dual type electronic switch like NPN and PNP transistor or N-channel and P-channel MOSFET, etc.

In the Fig. 6.5b, when d(t) i.e. digital baseband signal is 1, NPN transistor will remain in on state and PNP will be off. So, switch 1 will be directly connected to output. When d(t) is logic 0, the reverse action will occur i.e. PNP transistor will be on and NPN remains off. So now, switch 0 will be connected to output.

Therefore, it can obviously be inferred there is one switch/key which is responsible for shifting the amplitude from $A_c(1+m)$ to $A_c(1-m)$. Thus it is known as Amplitude Shift Keying.

Definition 6.1 *Amplitude Shift Keying (ASK) can be defined as 'A scheme of digital modulation where the instantaneous amplitude of the carrier is switched (keyed) between two levels controlled by the input digital stream.'*

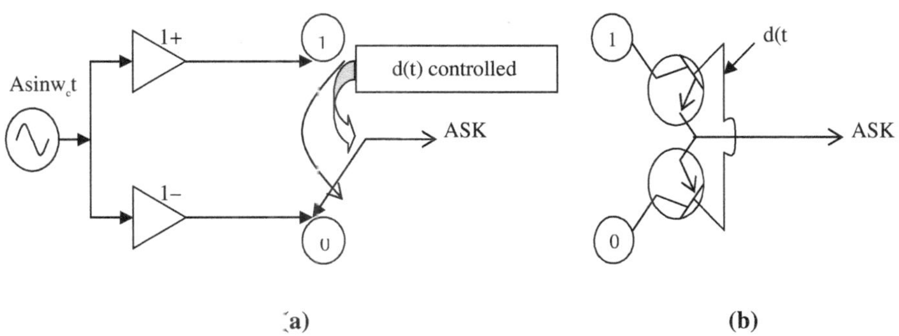

Fig. 6.5 (a) General ASK modulation, (b) Switching by NPN-PNP transistor (BJT) pair

As OOK is the practical ASK, therefore, OOK modulator is to be designed also in some elegant way.

By applying the incoming binary data (represented in unipolar form) and the sinusoid carrier to the two inputs of a product modulator (balanced modulator for DSB-SC generation), this ASK (OOK) signal can be generated (Fig. 6.6).

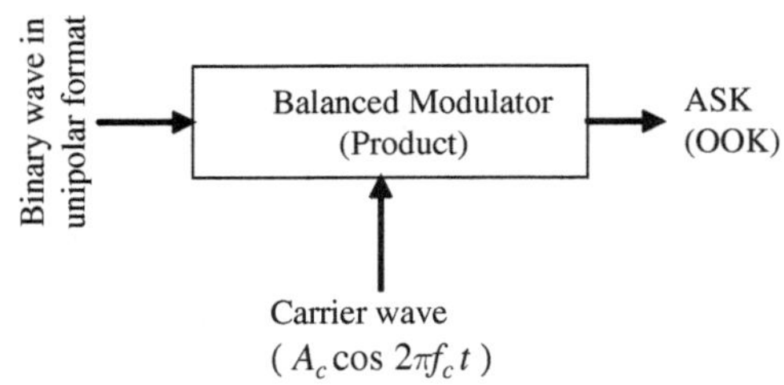

Fig. 6.6 Generation of binary ASK (OOK)

As DSB-SC modulation causes a shift of baseband signal spectrum to each side of the pilot frequency by an amount of the carrier frequency, here also ASK power spectral density (PSD) becomes similar to the shifted baseband signal spectrum by an amount $\pm f_c$. The justification is given in the following equation.

$$\begin{aligned}\varphi_{ASK}(t) &= d(t) \times A_c \cos \omega_c t \\ &= \frac{A_c}{2} \times d(t) \times \left\{e^{j\omega_c t} + e^{-j\omega_c t}\right\} \\ &= \frac{A_c}{2} \times \left[d(t)e^{i\omega_c t} + d(t)e^{-j\omega_c t}\right]\end{aligned} \tag{6.9}$$

By taking the Fourier transform of both sides,

$$\begin{aligned}\varphi_{ASK}(\omega) &= \frac{A_c}{2} F\left[d(t)e^{i\omega_c t} + d(t)e^{-j\omega_c t}\right] \\ &= \frac{A_c}{2}\left[D\left(\omega - \omega_c\right) + D\left(\omega + \omega_c\right)\right]\end{aligned} \tag{6.10}$$

where, $D(\omega)$ is the Fourier transform of d(t); as d(t) in a unit bit duration is a rect function (square type), therefore, $D(\omega)$ must be a *Sinc* type function. It is illustrated in Fig. 6.7. The spectrum of ASK signal [2] reveals that, it has an infinite bandwidth. For practical purposes, the bandwidth is often defined as the bandwidth of an ideal bandpass filter centered at f_c as because the central lobe of a *Sinc* contains 95% power. According to the above criteria, the bandwidth of ASK is $3/T_B$ Hz. The

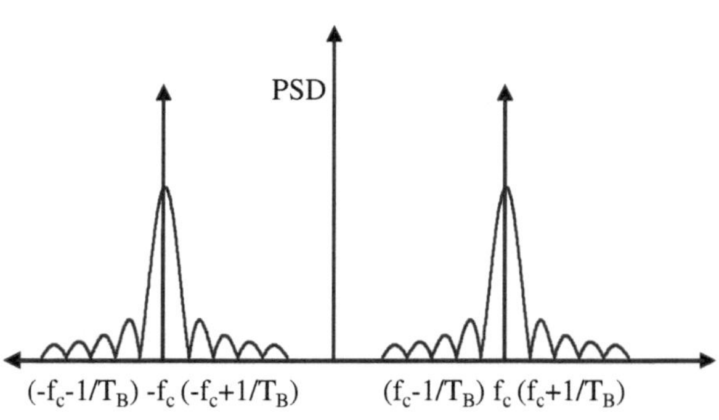

Fig. 6.7 Power spectral density of ASK signal

bandwidth of ASK can however be reduced by using smoothed version of digital pulse shape like Gaussian window.

6.2.3 Binary ASK Demodulator

Demodulation of binary ASK wave, i.e., binary baseband data can be achieved from the ASK waveform by using *coherent detector* [5] as shown in the Fig. 6.8. The detector consists of one product modulator, an integrator and a decision device. The incoming ASK signal is passed through a product modulation with a carrier of same phase and frequency to that of the carrier used in the transmitter section. Next, the integrator operates on the output of the multiplier for successive bit intervals and essentially performs a low-pass filtering action. The integrated (low-pass) signal is then passed through a decision making device. This is actually a comparator with a threshold. If the integrated output is higher than the threshold, the comparator makes a decision in favor of symbol 1; otherwise, the final output detected is symbol 0. Thus, a perfect digital signal is obtained at the output.

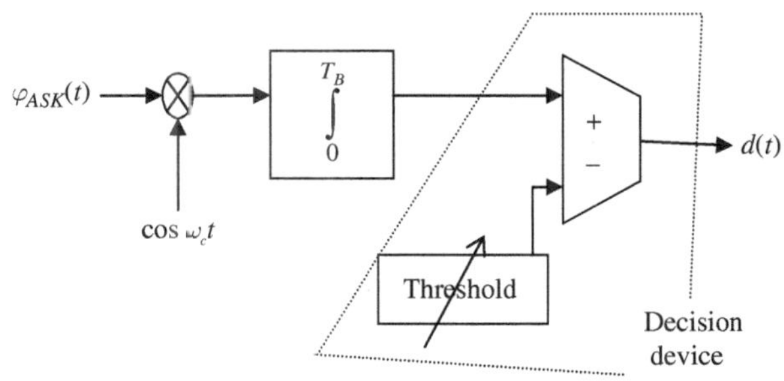

Fig. 6.8 Coherent detection of binary ASK signal

Binary ASK can also be demodulated *non-coherently* using envelope detector. Envelope detector essentially removes the complexity in synchronizing phase and frequency in case of coherent detection. The present scheme requires only a half wave rectification and low pass filtering at the receiver. The block diagram of a non-coherent receiver is shown in the Fig. 6.9.

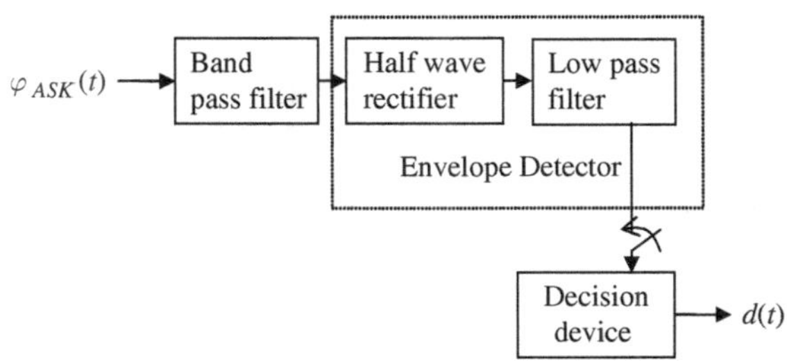

Fig. 6.9 Non-coherent detection of binary ASK signal

6.3 Frequency Shift Keying (FSK)

Frequency Shift Keying is analogous to Frequency Modulation (FM) in analog domain, where the slope of the time varying angle of the signal, i.e., instantaneous frequency of the carrier signal varies with respect to the baseband signal. In unipolar or bipolar formatting of the digital symbol, there exist two amplitude levels defined for two logic values one and zero. Therefore, two sinusoidal carrier waves of the same amplitude A_c but different frequencies f_H and f_L are used to represent the two symbols. The binary FSK wave may be represented as shown in Fig. 6.10.

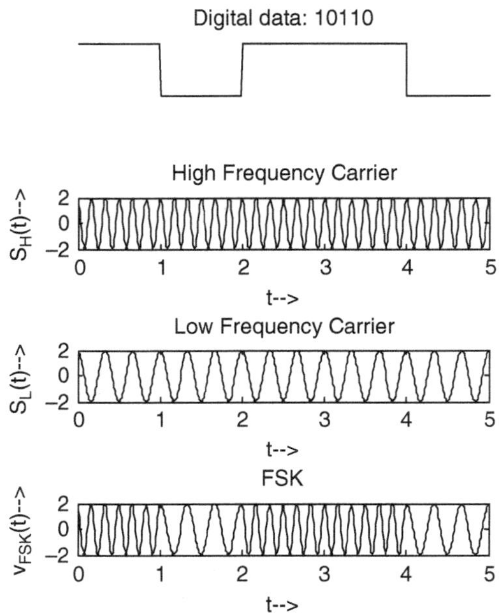

Fig. 6.10 FSK signal

6.3.1 Mathematical Model

According to the above discussion, the scheme of FSK modulation can therefore be described in a tabular form as shown in Table 6.2.

Table 6.2 FSK waveform definition

Data to be transmitted [d(t)]	FSK
Binary 1	$A_c \cos \omega_H t$
Binary 0	$A_c \cos \omega_L t$

The last column can be rewritten as,

$$\varphi_{FSK}(t) = A_c \cos \omega_H t \quad if \quad d(t) = Logic 1 \tag{6.11}$$

$$\varphi_{FSK}(t) = A_c \cos \omega_L t \quad if \quad d(t) = Logic0 \tag{6.12}$$

Therefore, it can obviously be inferred that, there is one *d(t)* controlled switch/key which is responsible for shifting the frequency of the carrier from f_H to f_L. Thus it is known as Frequency Shift Keying.

Definition 6.2 *Frequency Shift Keying (FSK) can be defined as 'A scheme of digital modulation where the instantaneous frequency of the carrier is switched (keyed) between two levels controlled by the input digital stream.'*

$$\text{RMS voltage of the carrier} = \frac{A_c}{\sqrt{2}}$$

$$\text{So, the normalized power (R=1), } P_S = \left(\frac{A_c}{\sqrt{2}}\right)^2 = \frac{A_c^2}{2}$$

$$\therefore A_c = \sqrt{2P_S}$$

Two make a concrete mathematical model, let's define two sinusoidal signals

$$S_H(t) = A_c \cos \omega_H t = \sqrt{2P_S} \cos \omega_H t \tag{6.13}$$

$$S_L(t) = A_c \cos \omega_L t = \sqrt{2P_S} \cos \omega_L t \tag{6.14}$$

Next, two properly shaped pulse train P_H and P_L are to be considered with the relation with d(t) according to the following Table 6.3.

Table 6.3 Data stream in terms of properly shaped pulses

Data to be transmitted [d(t)]	P_H	P_L
Binary 1	+1 V	0 V
Binary 0	0 V	+1 V

Therefore,

$$\varphi_{FSK}(t) = P_H S_H(t) + P_L S_L(t) \tag{6.15}$$

6.3.2 BFSK Modulator

From the mathematical model of binary FSK (BFSK), and Fig. 6.11, it's clear that, BFSK can be generated using two ASK modulators with inverted baseband and two carriers with frequencies f_H and f_L, as shown in the Fig. 6.13.

Two ASK signals must have the same bandwidth in order to generate a proper binary FSK signal. Another important point of the design is of course 'synchronization'. The delay introduced by the data inverter is to be taken care of by the

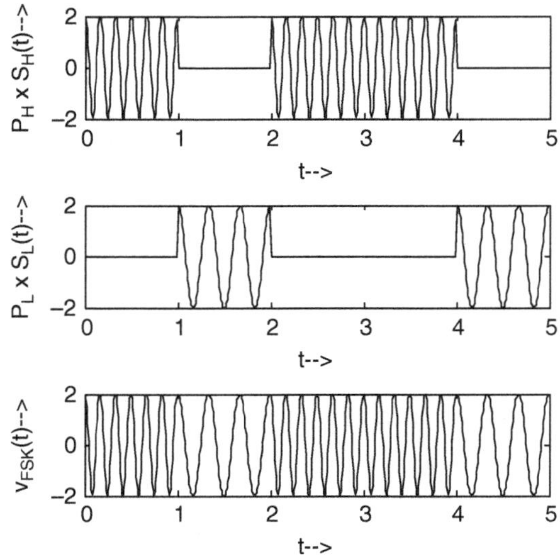

Fig. 6.11 FSK as combination of two ASK signals

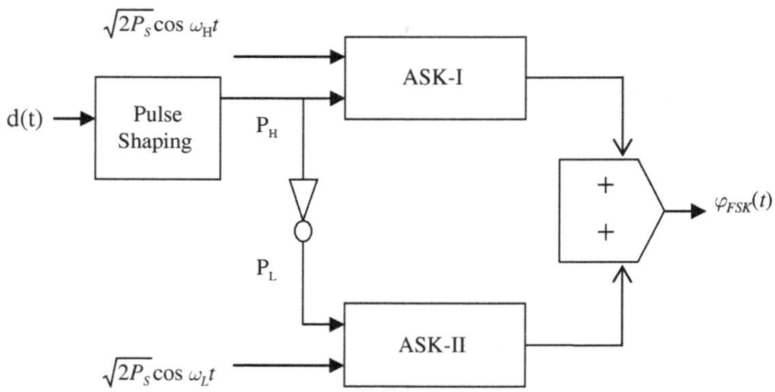

Fig. 6.12 BFSK generation using two ASK modulators

synchronizer ensuring that, the modulated waveforms (ASK-I and ASK-II) must arrive the data summer (Fig. 6.12) at the same instant.

From the spectrum of ASK, (Fig. 6.7), the PSD of BFSK also can be obtained. As FSK is the combination of two ASKs, the PSD of BFSK would also be summation of two *Sinc* functions. But the separation of the central lobe of the *Sinc* functions would greatly depend on the difference of frequencies f_H and f_L is comparison to the bit frequency $f_B = 1/T_B$. Different cases of this comparability are shown in the Fig. 6.13.

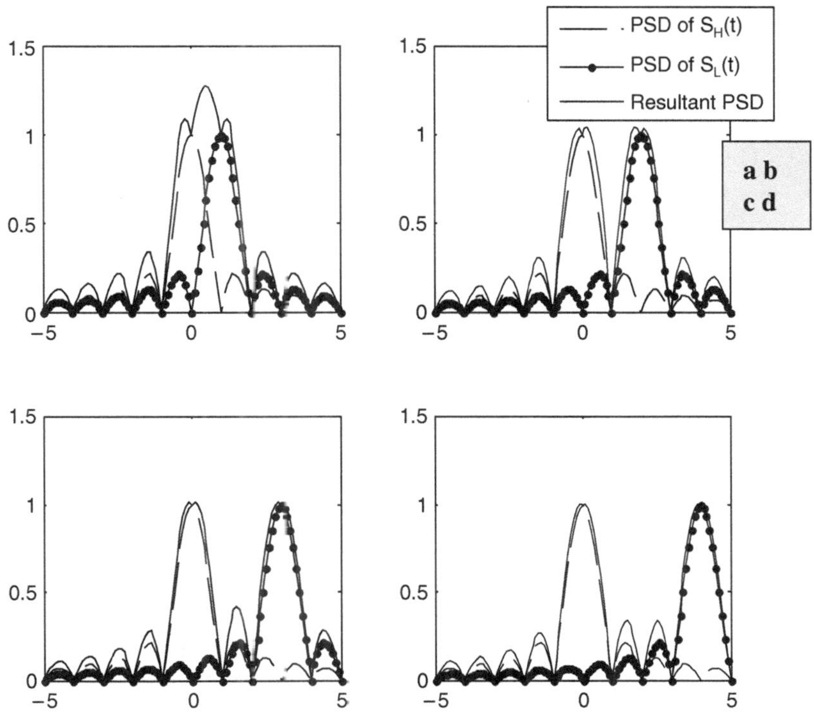

Fig. 6.13 PSD of BFSK: (**a**) $f_H - f_L < 2/T_B$, (**b**) $f_H - f_L = 2/T_B$, (**c**) $f_H - f_L > 2/T_B$, (**d**) $f_H - f_L >> 2/T_B$

6.3.3 FSK Demodulator

FSK signal can be detected using both coherent and non-coherent arrangements. As the binary FSK signal is generated using two carrier frequencies, the *coherent detector* consists of two correlators that are individually turned to two different carrier frequencies to represent symbol 1 and 0. A correlator is the combination of a multiplier followed by an integrator (LPF). The same BFSK signal is multiplied by two different carrier signals and the passed separately through the LPF. The outputs of the LPF must be inverted type, i.e., if one is upper going, the other must be lower going. Next, the outputs are fed to the decision deice which is essentially a comparator. If the output produced in the upper path is more than the output produced in the lower path, the magnitude comparator will decide in favor of symbol 1. Similarly, if the output of the upper path is less than that of the lower path, the comparator will decide in favor of symbol 0. This type of digital communication receivers are also called as correlator receiver (Fig. 6.14).

As in case of ASK, binary FSK also can be demodulated *non-coherently* using envelope detector. The received BFSK signal is applied to the bank of two band pass filters (parallel BPF combination) with central pass frequencies f_H and f_L. Each

Fig. 6.14 Coherent detection of binary FSK signal

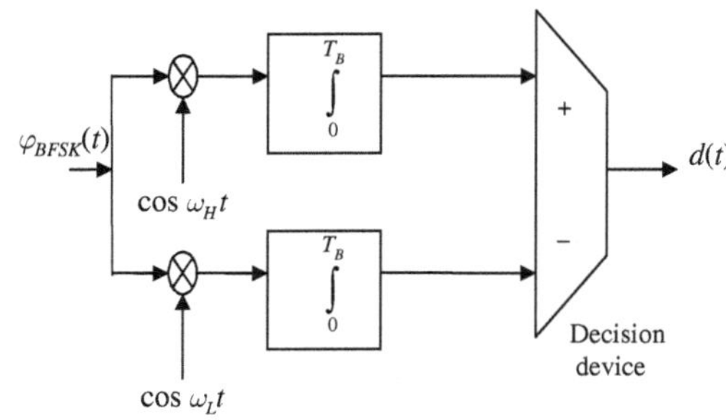

filter is followed by an envelope detector. It is not at all difficult to understand that, the output of the envelope detectors would be P_H and P_L, i.e., the outputs would be inverted to each other. Therefore, the comparator as shown in Fig. 6.15 would decide in favor of symbol 1 is $P_H > P_L$, and decision would be made in favor of symbol 0, otherwise.

Fig. 6.15 Non-coherent detection of binary FSK signal

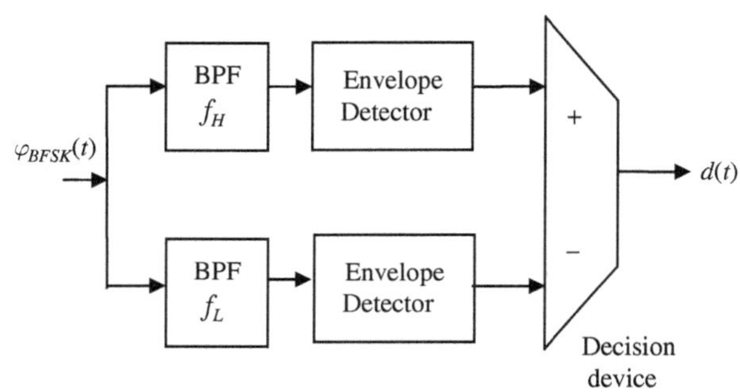

6.4 Binary Phase Shift Keying (BPSK)

Phase Shift Keying is analogous to Phase Modulation (PM) in analog domain, where the slope of the time varying angle of the signal, i.e., instantaneous phase of the carrier signal varies with respect to the baseband signal. Here, a fixed frequency sinusoid is used to represent two symbols, 1 and 0. To differentiate the representing waveforms, there exists a phase difference of 180°.

The scheme of modulation is as follows

$$\varphi_{BPSK}(t) = A_c \cos \omega_c t \qquad if \quad d(t) = Logic1 \tag{6.16}$$

$$\varphi_{BPSK}(t) = A_c \cos(\omega_c t + \pi) = -A_c \cos \omega_c t \qquad if \quad d(t) = Logic0 \tag{6.17}$$

By employing this scheme for digital modulation of binary data d(t)=10110001, the following waveform is obtained (Fig. 6.16).

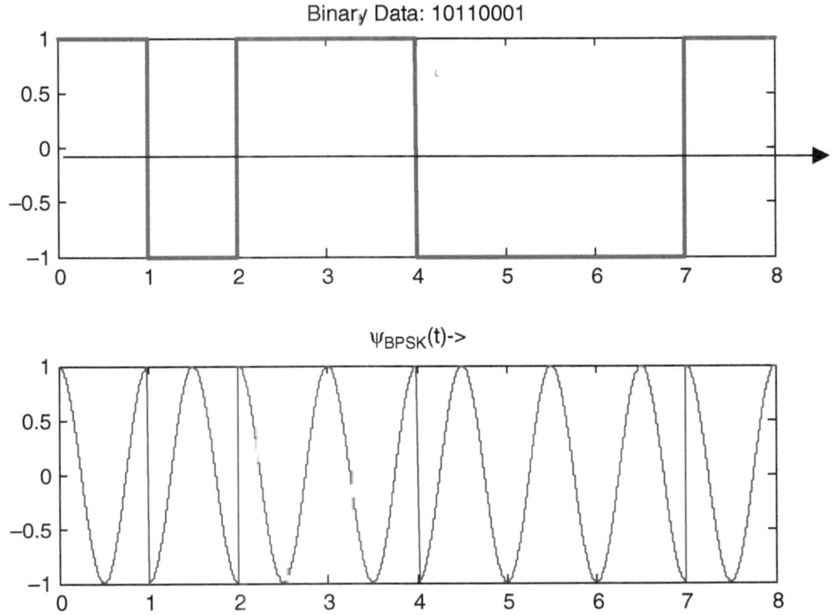

Fig. 6.16 Binary PSK waveform

6.4.1 Mathematical Model

According to the above discussion, the scheme of BPSK modulation can therefore be described in a tabular form as in Table 6.4.

Table 6.4 BPSK waveform definition

Data to be transmitted [d(t)]	Analog equivalent definition	ASK
Binary 1	+1 V	$A_c \cos \omega_c t$
Binary 0	−1 V	$-A_c \cos \omega_c t$

Combining the two rows of the last column of Table 6.4, the expression of BPSK is as follows:

$$\varphi_{BPSK}(t) = A_c d(t) \cos \omega_c t \tag{6.18}$$

Mathematically, the scheme can therefore be modeled as the Eq. (6.18), where, d(t) is in bipolar format.

In other words, BPSK is the combination of two sinusoids one in phase and other 180° out of phase. The scheme can be viewed as shown in the Fig. 6.17. There are two terminals (0) and (1) which are the responsible outputs for digitally modulated waveforms by symbol 1 and 0 respectively. A d(t) controlled switch is there to shift

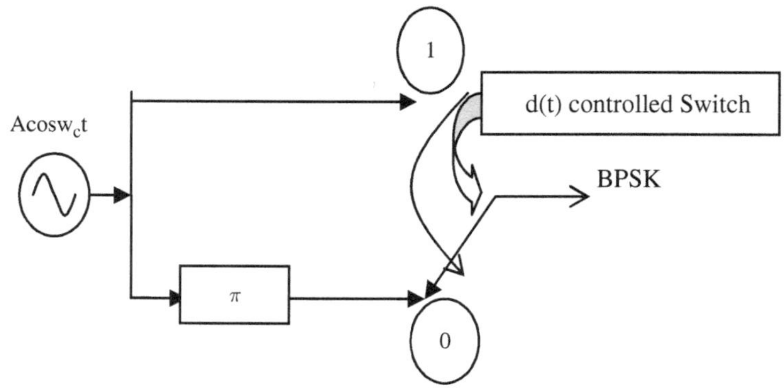

Fig. 6.17 Concept of phase switching in BFSK signal

the phase of the signal from 0 to 180° and vice versa. From this idea, the BFSK can be defined easily (Fig. 6.17).

Definition 6.3 *Phase Shift Keying (PSK) can be defined as 'A scheme of digital modulation where the instantaneous phase/angle of the carrier is switched (keyed) between two values (0 and π) controlled by the input digital stream.'*

6.4.2 BPSK Modulator

From the discussion of the previous section and from Eq. (6.16), binary PSK waveform can be viewed as the product of binary data d(t) and sinusoid carrier. As Fig. 6.16 shows the negativity of the sinusoid when represented for binary data bit zero, the product modulation (Fig. 6.18) must be done between sinusoid carrier and digital data d(t) in *bipolar format.*[1]

Fig. 6.18 Generation of binary PSK

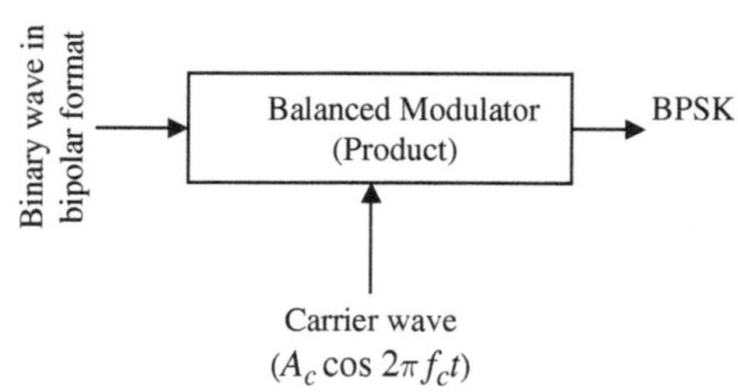

Power Spectral Density (PSD) of BFSK is therefore very similar to the PSD of ASK. The only difference is that the former does not have an impulse at the

[1] It is an interesting observation that, to generate ASK also, same product modulator is used but the only difference is in the format of d(t). In ASK modulator unipolar d(t) is used (Fig. 6.6) whereas in PSK, bipolar d(t) is used (Fig. 6.18).

carrier frequency. The bandwidth requirement is $2/T_B$. The most significant difference between ASK and BPSK is that, ASK is linear modulation whereas PSK is non-linear modulation technique. Therefore, PSK has a superior performance [3] over ASK in a noisy environment for a given peak power (Fig. 6.19).

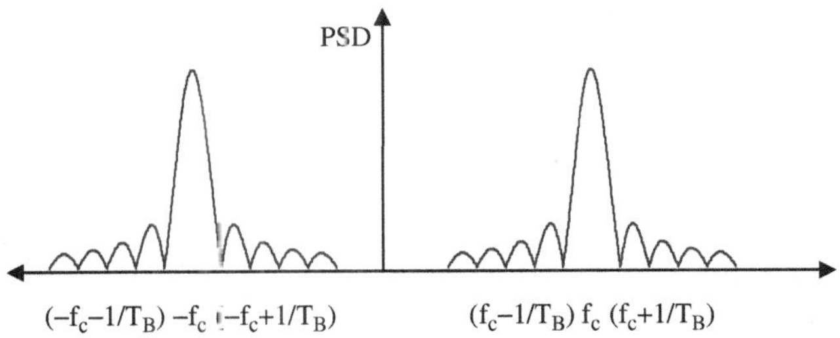

Fig. 6.19 PSD of binary PSK waveform

6.4.3 BPSK Demodulator

The *coherent detection* of BPSK is not very different from ASK detection (as shown in Fig. 6.8). The detector consists of a product modulator synchronized with a locally sinusoid carrier, an integrator that's operates on the multiplier output for successive bit interval and a decision device that compares the integrated (LPF) output with a predefined threshold. Though this coherent detector suffers from the problem of synchronization both in pulse and timing, PSK cannot e detected using non-coherent detector. This is because of the fact that, the envelopes extracted from the waveform representing symbol 1 and symbol 0 are not indifferent. To avoid the necessity of synchronization, a modified PSK can be employed. In *Differential Phase Shift Keying* there is no need of synchronous carrier to demodulate the PSK.

6.5 Differential Phase Shift Keying (DPSK)

Differential Phase Shift Keying necessarily deals with the signal transition from 0 to 1 and from 1 to 0. In this particular scheme of modulation, phase synchronization is not at all required, as BPSK. For example, the symbol 0 may be used to represent transition in the baseband binary sequence (with respect to the previous encoded bit) and symbol 1 to represent no transition.

6.5.1 DPSK Modulator

The block diagram of DPSK modulator is shown in Fig. 6.20. The binary data stream d(t) is applied to the input of the DPSK encoder. The output is then applied to a

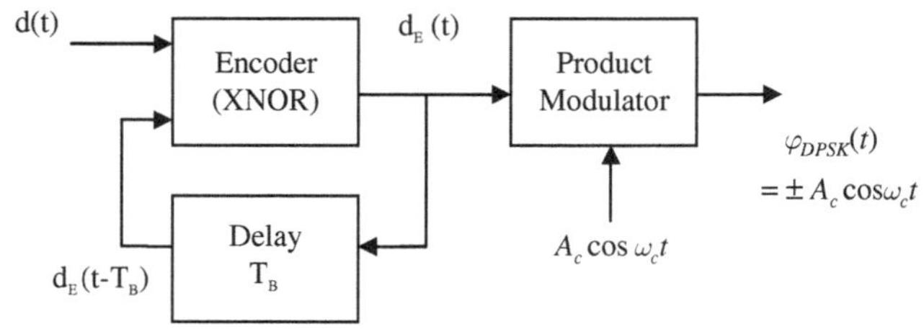

Fig. 6.20 DPSK modulator

product modulator (DSB-SC). A constant amplitude sinusoid with fixed frequency is applied to the other input of the product modulator. The fashion of generation of differential bits is depicted in the Table 6.5 for data sequence d(t)=10110001. In the illustration, an extra bit, i.e. symbol 1 is added as an initial bit.[2] This is to sustain the continuity of data encoding. The phase of the generated DPSK signal is also shown in the Table 6.5. It is to be noted from the Fig.6.20 that, the input binary bit stream is denoted as d(t) and differentially encoded bit stream is denoted as $d_E(t)$.

Table 6.5 Differential encoded sequences with phase

d(t)		1	0	1	1	0	0	0	1
$d_E(t)$ **[XNOR]**	1*	1	0	0	0	1	0	1	1
Phase of DPSK	0	0	π	π	π	0	π	0	0
Shifted $d_E(t)$		1*	1	0	0	0	1	0	1
Phase of $d_E(t)$		0	0	π	π	π	0	π	0
Phase comparison		+	-	+	+	-	-	-	+
Detected binary data Sequence $\hat{d}(t)$		1	0	1	1	0	0	0	1
***Arbitrary starting reference bit**									

[2]It is to be noted that, Table 6.5 can also be reconstructed taking the arbitrary initial bit as 0. There will be absolutely no problem in differential encoding and decoding.

6.5.2 DPSK Demodulator

At the receiver side, the DPSK waveform is applied to one input of the multiplier alongwith its one bit interval delayed version. The delayed DPSK is shown in the 4th row of the Table 6.5. The output of the difference is proportional to cos ϕ where ϕ is the difference between the carrier phase angle of the received DPSK signal and its delayed version, measured in the same bit interval. The phase difference between the two waveforms for each bit interval is used to determine the sign of the phase comparator output. The phase comparator is nothing but the combination of an analog multiplier and a low pass filter (or integrator). When $\phi = 0$, the integrator output is positive whereas when $\phi = \pi$, the output is negative. Hence, a zero crossing detector can be used as the decision device to detect the binary symbol accurately (Fig. 6.21).

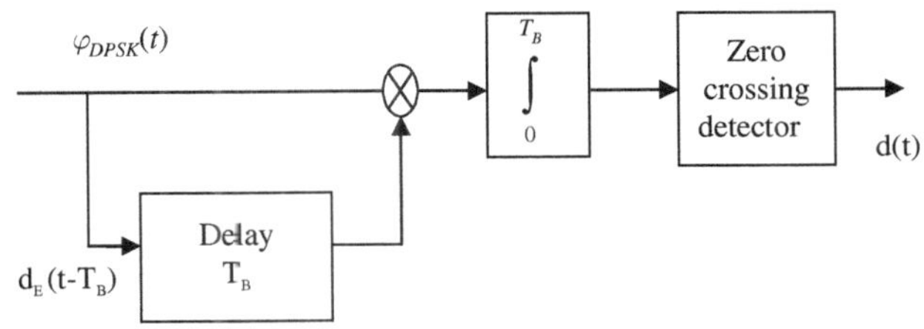

Fig. 6.21 DPSK demodulator

6.6 Quadrature Phase Shift Keying (QPSK)

Spectrally efficient modulation technique [3] is used essentially for maximizing the transmission bandwidth efficiency. The phrase 'spectrally efficient modulation' implies, a constant bandwidth ($2/T_B$) is made usable by two symbols instead of a single bit like BPSK. The only technique which can be applied to get this spectral efficiency is modulation the pair of bits (di-Bit) through a pair of carriers of same frequency and quadrature in phase. The phase quadrature modulation prevents the modulated signal to superimpose, i.e., the bandwidth is usable by the representations of two symbols. In BPSK, single bit was in consideration at a time. To represent the modulated waveform, one in phase and another out of phase component was taken into account as shown in the Fig. 6.22a. But, in case of QPSK, at a time two bits (di-bit) are considered. Total number of combinations formed by 2 bits is 4. Therefore, in the constellation diagram, 4 points are to be indicated placed in equidistance manner as shown in Fig. 6.22b.

6.6.1 Mathematical Model

From the discussion above, a generalized expression of QPSK can be modeled as follows:

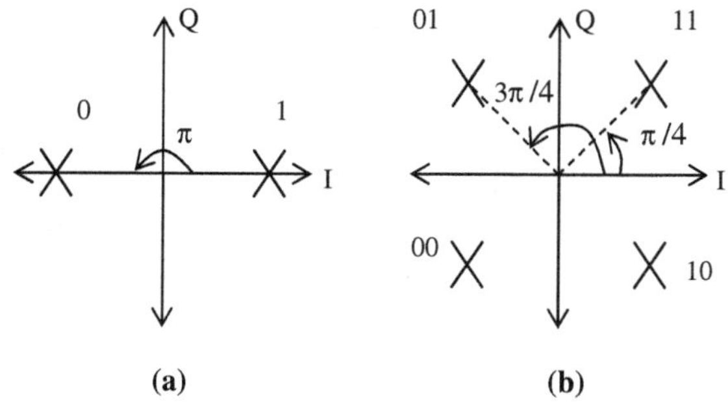

Fig. 6.22 I-Q diagram of (**a**) BPSK and (**b**) QPSK

$$\varphi_{QPSK}(t) = A_c \cos(\omega_c t + \phi_n) \tag{6.19}$$

where, $\phi_n = (2n+1)(\pi/4)$, for n=0,1,2,3
For n=0, taking $A_c = 1$,

$$\begin{aligned}\varphi_{QPSK}(t) &= \cos(\omega_c t + \pi/4) \\ &= \cos\omega_c t \cos(\pi/4) - \sin\omega_c t \sin(\pi/4) \\ &= \frac{1}{\sqrt{2}}(\cos\omega_c t - \sin\omega_c t)\end{aligned} \tag{6.20}$$

From Eq. (6.20), $\cos\omega_c t$ and $-\sin\omega_c t$ are in-phase and quadrature phase carrier basis, termed as $\xi_I(t)$ and $\xi_Q(t)$ respectively. Therefore, Eq. (6.20) can be re-written as

$$\varphi_{QPSK}(t) = \frac{1}{\sqrt{2}}\left\{(+1)\,\xi_I(t) + (+1)\,\xi_Q(t)\right\} \tag{6.21}$$

The 1st step towards realizing QPSK modulation is partitioning of pulses. The original data stream consisting of bipolar pulses, i.e., $d_k(t) = +1$ or -1, representing binary 1 and 0 respectively is expressed as $d_k(t) = d_0, d_1, d_2, \ldots$

This pulse stream is partitioned into an in-phase stream $d_I(t)$, and a quadrature stream $d_Q(t)$ taking odd and even bits separately.

$$d_I(t) = d_0, d_2, d_4, \ldots . \text{ (Even bits)} \tag{6.22}$$

$$d_Q(t) = d_1, d_3, d_5, \ldots . \text{ (Odd bits)} \tag{6.23}$$

From the Fig. 6.23, it's clear that, bit rate of $d_I(t)$ and $d_Q(t)$ each is actually the half of the bit rate of $d_k(t)$.

Therefore, a convenient orthogonal realization [4] of the QPSK waveform is achieved by amplitude modulation by the in-phase and quadric-phase data partitions into cosine and sine carrier waves, respectively. Hence, the generalized

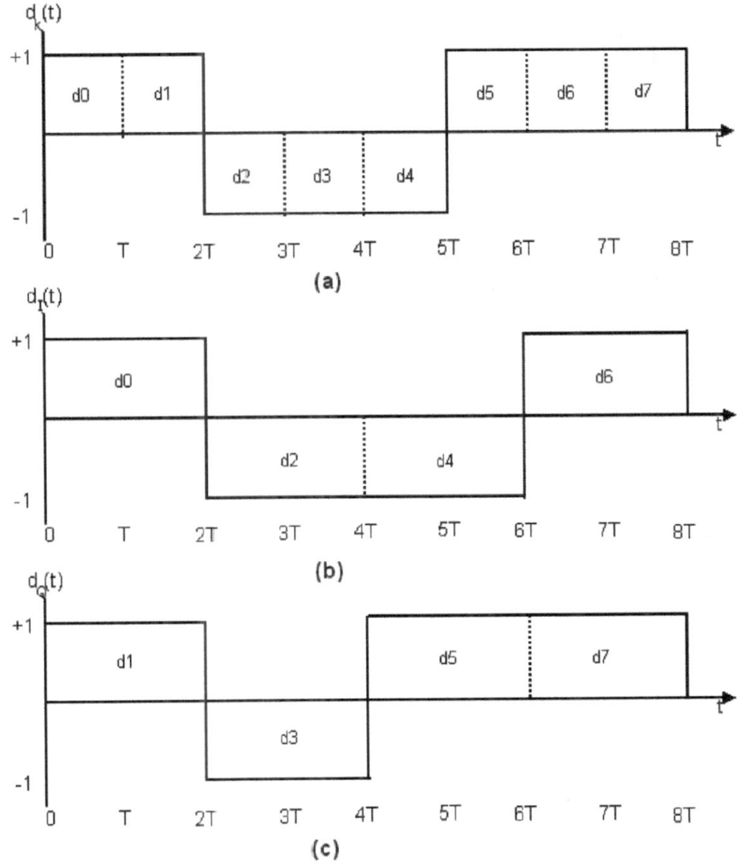

Fig. 6.23 (**a**) Original data stream, (**b**) partitioned in-phase data stream, (**c**) partitioned quadrature phase data stream

mathematical model is as follows:

$$\begin{aligned}\varphi_{QPSK}(t) &= \frac{1}{\sqrt{2}}\left(d_I(t)\cos\omega_c t - d_Q(t)\sin\omega_c t\right)\\ &= \frac{1}{\sqrt{2}}\left(d_I(t)\xi_I(t) + d_Q(t)\xi_Q(t)\right)\end{aligned} \tag{6.24}$$

Hence, the 4 di-Bit arrangements can be formed taking the help of the Eqs. (6.24) and (6.19).

The derived di-bit is indicated into the I-Q diagram of QPSK (Fig. 6.22b). It's interesting to note that, one bit (either LSB or MSB) is being changed at a time when the angle is being increased in steps of $\pi/4$ (Table. 6.6). This is indeed the form of Gray code.

Table 6.6 QPSK waveforms and di-bit representation

n	$\varphi_{QPSK}(t)$ from Eq. (6.19)	QPSK waveform	$\varphi_{QPSk}(t)$ from Eq. (6.24)	di-Bit $d_I(t)$	$d_q(t)$
0	$A_c \cos(\omega_c t + \pi/4)$	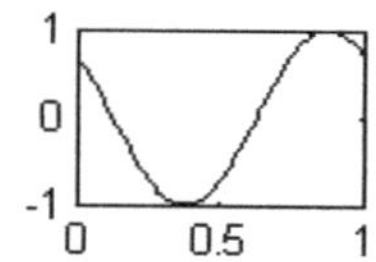	$\frac{1}{\sqrt{2}}((+1)\xi_I(t) + (+1)\xi_Q(t))$ as $\cos(\pi/4) = 1/\sqrt{2}, \sin(\pi/4) = 1/\sqrt{2}$	1	1
1	$A_c \cos(\omega_c t + 3\pi/4)$	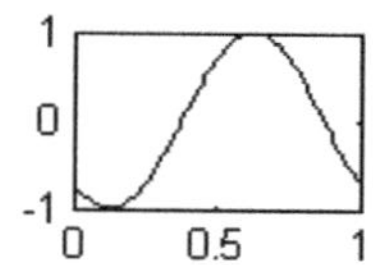	$\frac{1}{\sqrt{2}}((-1)\xi_I(t) + (+1)\xi_Q(t))$ as $\cos(3\pi/4) = -1/\sqrt{2}, \sin(3\pi/4) = 1/\sqrt{2}$	0	1
2	$A_c \cos(\omega_c t + 5\pi/4)$	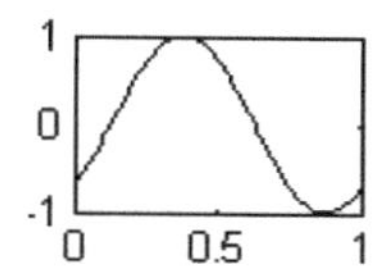	$\frac{1}{\sqrt{2}}((-1)\xi_I(t) + (-1)\xi_Q(t))$ as $\cos(5\pi/4) = -1/\sqrt{2}, \sin(5\pi/4) = 1/\sqrt{2}$	0	0
3	$A_c \cos(\omega_c t + 7\pi/4)$	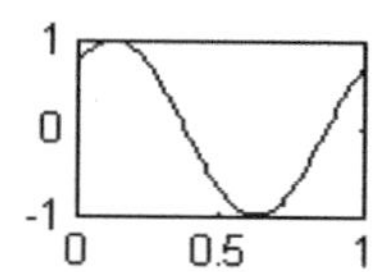	$\frac{1}{\sqrt{2}}((+1)\xi_I(t) + (-1)\xi_Q(t))$ as $\cos(7\pi/4) = 1/\sqrt{2}, \sin(7\pi/4) = 1/\sqrt{2}$	1	0

6.6.2 QPSK Modulator

The mathematical model always helps us to generate the modulated waveform in a convenient way. As suggested by the model, 1st the incoming bits are being partitioned into odd and even (in-phase and quadrature phase) bits using a serial to parallel converter working in frequency 0.5 r_B, where r_B=bit rate=$1/T_B$. The odd bits $d_I(t)$ are modulated using in phase carrier component $A_c \cos \omega_c t$, as shown in Fig. 6.24. By a phase shifter of 90°, the quadrature phase carrier is also generated and used for modulation of even bits $d_Q(t)$. The summation of these two modulated waveform is QPSK.

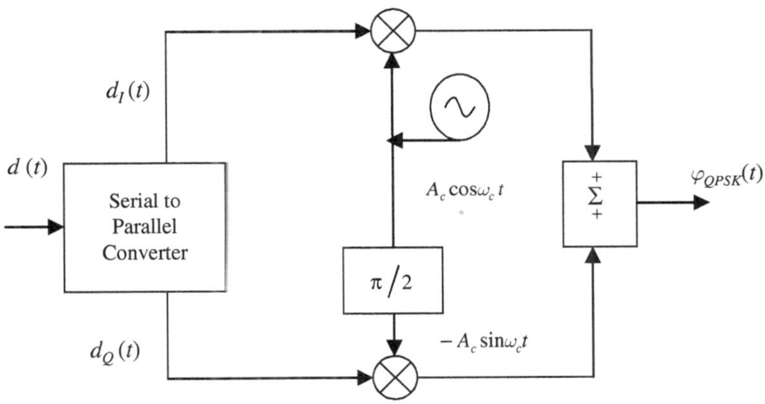

Fig. 6.24 QPSK modulator

The QPSK waveform for binary data 10110001 is shown in Fig. 6.25. The figure clearly shows how a pair of bit is taken for modulation. The frequency of carrier is exactly equal to the bit rate ($1/T_B$).

6.6.3 QPSK Demodulator

The QPSK signal can be detected using an arrangement of parallel combination of a pair of correlator as show in the Fig. 6.26. The upper correlator computes the cosine of the carrier phase whereas the lower one computes the sine of the carrier phase. The sign of the correlators are then computed using two different decision devices. The parallel to serial converter interleaves the decision made by the outputs of the two correlators, i.e., the in-phase and quadrature phase channels of the receiver and henceforth reconstructs the information data stream. In absence of bit reversal in channels these pair of transmitter-receiver (modulator–demodulator) results in a successful communication, i.e., the transmitted bits can be received un-erroneously at the receiver side.

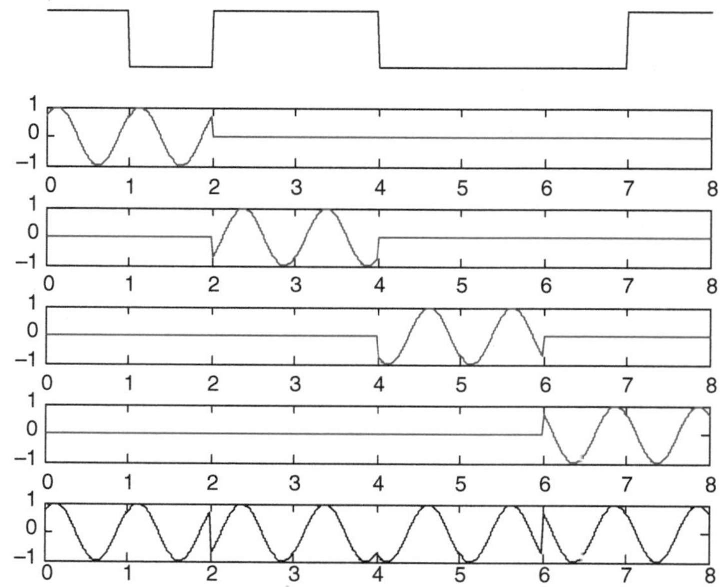

Fig. 6.25 QPSK waveform for binary data 10110001

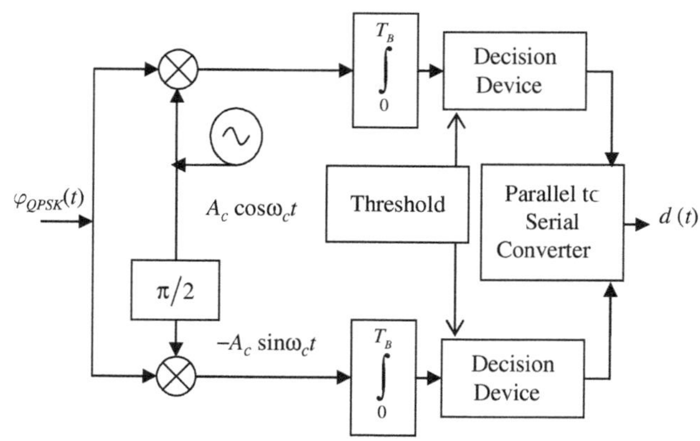

Fig. 6.26 QPSK coherent demodulator

6.6.4 *Offset QPSK (OQPSK)*

Offset QPSK modulation [7] is a specially designed QPSK to avoid out-of-band interference. The difference between QPSK and OQPSK is only in the *alignment* of the two baseband signals. As shown in Fig. 6.27, the duration of each original pulse is T. Hence, when we consider di-bit (in the partitioned space), the duration of each pulse becomes 2T. In standard QPSK, the odd and even pulse stream are both transmitted at a rate of 1/2T bit/s and are synchronously aligned, such that their transition coincide (Fig. 6.23). In OQPSK, there is the same data stream partitioning and orthogonal transmission; but, the timing of the pulse streams $d_I(t)$ and $d_Q(t)$ is shifted such that the alignment of the two streams is offset by T. The offset is clearly depicted in the Fig. 6.27.

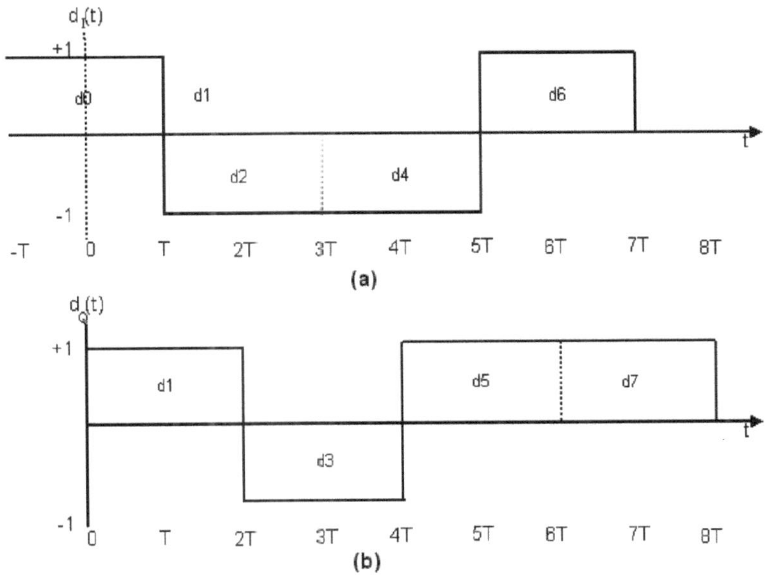

Fig. 6.27 Offset QPSK data stream partitioning with one bit interval time offset

In standard QPSK, the in-phase (odd) and quadrature phase (even) bit stream are transmitted at the rate of 1/2T bit/s and are synchronously aligned, such that their transition coincide. But in OQPSK, the even and odd bit streams are aligned in shifted manner. Due to coincident alignment in QPSK, the modulated carrier phase can change only once every $2T_B$ interval. During the next $2T_B$ interval, if only one of the pulse streams change sign, a phase shift of 90° occurs. For change in both the streams, a phase shift of 180° results in.

To reduce the occurrence of spectral side lobes, when the QPSK waveform is filtered in, the resulting waveform would not be a constant envelope signal. The occasional phase shift of 180° would cause the envelope to go to zero momentarily. For this reason, the undesirable side lobes frequencies are restored which in tern interfere with the nearby channels.

On the other hand, in OQPSK only one component of bit stream (either odd or even) is allowed to be changed. Therefore, change of phase is either 0° or 90°, but not 180°. In Fig. 6.28, comparison between the waveforms of QPSK and OQPSK is presented. When OQPSK signal is band limited, the envelope will not go to zero. Hence, the high-frequency components associated with the collapse of the envelope are not reinforced. Thus out of band interference is avoided.

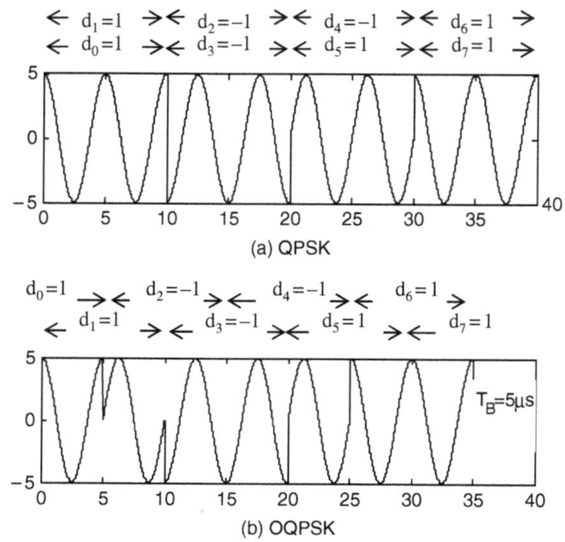

Fig. 6.28 (**a**) QPSK waveform, (**b**) Offset QPSK waveform

6.7 Minimum Shift Keying (MSK)

QPSK scheme is affected with the obvious discontinuous phase transition. To avoid the discontinuous phase transition problem, there was a motivation for designing continuous phase modulation (CPM) schemes [1]. Minimum shift keying (MSK) is one such scheme. MSK can be viewed as a special case of continuous phase frequency shift keying (CPFSK), expressed as:

$$s(t) = \cos\left[2\pi\left(f_0 + \frac{d_k}{4T}\right)t + x_k\right] \qquad ; kT < t < (k+1)T \tag{6.25}$$

where, f_0 is the carrier frequency, $d_k = \pm 1$ represents the bipolar data being transmitted at a rate $R = 1/T$, and x_k is a phase constant which is valid over the kth binary data interval. It is to be noted that, for $d_k = 1$, the frequency transmitted is $(f_0 + 1/4T)$, and for $d_k = -1$, the frequency transmitted is $(f_0 - 1/4T)$. The tone spacing in MSK is thus one-half that employed for non-coherently demodulated orthogonal *FSK*, giving rise the name minimum shift keying. During each data interval T, the value of x_k is constant, i.e., $x_k = 0$ or π, determined by the phase of the modulated waveform be continuous at $t = kT$. This requirement results in the following recursive phase constraint for x_k :

$$x_k = \left[x_{k-1} + \frac{\pi k}{2}(d_{k-1} - d_k)\right] \text{modulo } 2\pi \tag{6.26}$$

Equation (6.25) can be expressed in a quadrature representation as expressed in QPSK,

$$s(t) = a_k \cos \frac{\pi t}{2T} \cos 2\pi f_0 t - b_k \sin \frac{\pi t}{2T} \sin 2\pi f_0 t \quad ; kT < t < (k+1)T \tag{6.27}$$

where,

$$\begin{aligned} a_k &= \cos x_k = \pm 1 \\ b_k &= d_k \cos x_k = \pm 1 \end{aligned} \tag{6.28}$$

The in-phase (I), and quadrature phase (Q) components are identified as $a_k \cos \pi t/2T \cos 2\pi f_0 t$ and $b_k \sin \pi t/2T \sin 2\pi f_0 t$ respectively, where $\cos \pi t/2T$ and $\sin \pi t/2T$ are the *sinusoidal symbol weighing*; $\cos 2\pi f_0 t$ and $\sin 2\pi f_0 t$ are the in-phase and quadric-phase carrier terms. Since the source data d_k changes every T seconds, and a_k, b_k both are data dependent terms, it might appear that, a_k and b_k terms can change every T seconds. However, because of the continuous phase constraint, the a_k and b_k term can only change value at the zero crossings of $\cos \pi t/2T$ and $\sin \pi t/2T$ respectively. Therefore, the symbol weighing in either the I-channel or Q-channel is a half-cycle sinusoid pulse of duration 2T seconds with alternate sign.

As shown in the equation Eq. (6.26), x_k is a function of difference between the present and prior data bits, a_k and b_k can be viewed as *differentially encoded* components of the source data. Thus, Eq. (6.27), can be rewritten as:

$$s(t) = d_I(t) \cos \frac{\pi t}{2T} \cos 2\pi f_0 t + d_Q(t) \sin \frac{\pi t}{2T} \sin 2\pi f_0 t \tag{6.29}$$

where, $d_I(t)$ and $d_Q(t)$ are the in-phase and quadric-phase data stream.

The Fig. 6.29 shows the sinusoid weighing of the I-channel and Q-channel pulses. From the Fig. 6.29, we can infer about the following properties of the MSK modulation as:

1. The waveform s(t) has a constant envelope;
2. There is a phase continuity in the carrier at the bit transition instant;
3. The waveform can be regarded as an FSK waveform with signaling frequencies $(f_0 + 1/4T)$ and $(f_0 - 1/4T)$.

Therefore, the minimum tone separation required for MSK modulation is

$$\left(f_0 + \frac{1}{4T}\right) - \left(f_0 - \frac{1}{4T}\right) = \frac{1}{2T} \tag{6.30}$$

which is equal to the half of the bit rate.

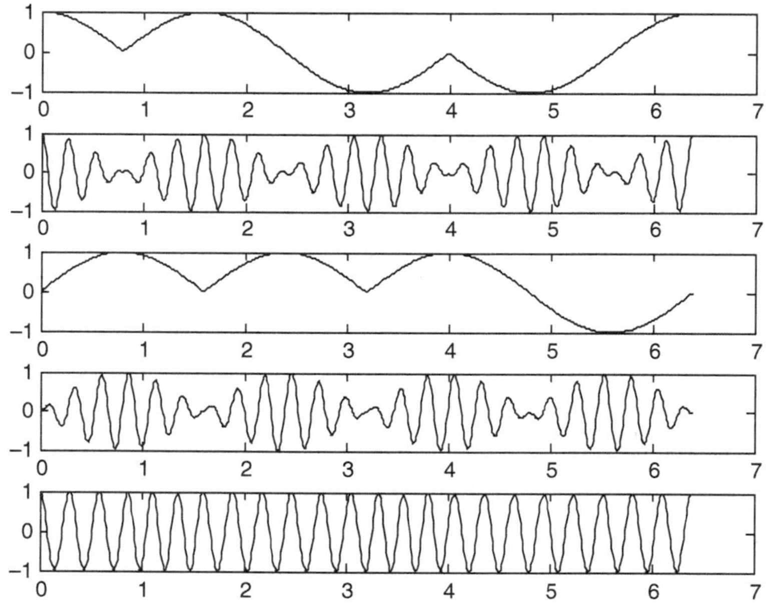

Fig. 6.29 Minimum shift keying

6.8 Probability of Error for Different Modulation Schemes

In Chap.3, the calculation of probability of error for baseband communication is derived in detail. In baseband communication the symbol 0 and 1 are represented using some digital formats. No modulation is employed there. The difference between the calculation of probability of error of baseband communication and that of modulation schemes lies there. In modulation, symbol 1 and 0 are represented in terms of different waveforms. In this section, the probability of error is calculated for different modulation schemes when passed through a Matched filter.

6.8.1 Probability of Error in ASK

In ASK, signal to represent symbol one is $S_1(t) = A\cos\omega_c t$ and signal to represent binary digit 0 is $S_0(t) = 0$.

Thus, from Eq. (3.18), the difference signal

$$p_i(t) = S_1(t) - S_0(t) = A\cos\omega_c t$$

When the ASK signal is passed through a matched filter (i.e. optimum filter with AWGN), the optimization parameter is maximized to

$$\gamma^2{}_{max} = \frac{2}{\eta}\int_0^{T_B} \{S_1(t) - S_0(t)\}^2\, dt \text{ from Eq. (3.44)}$$

$$= \frac{2}{\eta}\int_0^{T_B} A^2 \cos^2 \omega_c t dt \qquad (6.31)$$

$$= \frac{A^2 T_B}{\eta}$$

Hence, from Eq. (3.46), probability of error is

$$P_{e\ ASK} = \frac{1}{2} erfc\left[\frac{\gamma^2}{8}\right]^{1/2}$$

$$= \frac{1}{2} erfc\left[\frac{A^2 T_B}{8\eta}\right]^{1/2} \qquad (6.32)$$

$$= \frac{1}{2} erfc\left[\frac{E_S}{4\eta}\right]^{1/2}$$

$$\left[\because E_S = \left(\frac{A}{\sqrt{2}}\right)^2 T_B\right]$$

6.8.2 Probability of Error in FSK

In FSK, signal to represent symbol one is $S_1(t) = A\cos(\omega_c + \Omega)t$ and signal to represent binary digit 0 is $S_0(t) = A\cos(\omega_c - \Omega)t$, where, $\omega_H = (\omega_c + \Omega)$ and $\omega_L = (\omega_c - \Omega)$ comparing with Eqs. (6.11) and (6.12).

Therefore, the difference signal

$$p_i(t) = S_1(t) - S_0(t) = A\{\cos(\omega_c + \Omega)t - \cos(\omega_c - \Omega)t\}$$

Now, the maximized optimization parameter is

$$\gamma^2{}_{max} = \frac{2}{\eta}\int_0^{T_B} \{S_1(t) - S_0(t)\}^2\, dt$$

$$= \frac{2}{\eta}\int_0^{T_B} A^2\{\cos(\omega_c + \Omega)t - \cos(\omega_c - \Omega)t\}^2\, dt$$

$$= \frac{2A^2T_B}{\eta}\left[1 - \frac{\sin 2\Omega T_B}{2\Omega T_B} + \frac{1}{2}\frac{\sin 2\left(\omega_c + \Omega\right)T_B}{2\left(\omega_c + \Omega\right)T_B} - \frac{1}{2}\frac{\sin 2\left(\omega_c - \Omega\right)T_B}{2\left(\omega_c - \Omega\right)T_B} - \frac{\sin 2\omega_c T_B}{2\omega_c T_B}\right] \tag{6.33}$$

With the assumptions that, $\omega_c T_B >> 1$ and $\omega_c >> \Omega$ for practical systems, the last three terms of Eq. (6.33) are neglected. Hence,

$$\gamma^2{}_{\max} = \frac{2A^2T_B}{\eta}\left[1 - \frac{\sin 2\Omega T_B}{2\Omega T_B}\right] \tag{6.34}$$

$\gamma^2{}_{\max}$ will attain maximum value, when $2\Omega T_B = 3\pi/2$. Putting the value,

$$\gamma^2{}_{\max} = 2.42\frac{A^2T_B}{\eta} \tag{6.35}$$

Hence,

$$P_e|_{FSK} = \frac{1}{2}erfc\left[\frac{\gamma^2{}_{\max}}{8}\right]^{1/2} = \frac{1}{2}erfc\left[\frac{0.3A^2T_B}{\eta}\right]^{1/2} = \frac{1}{2}erfc\left[\frac{0.6E_S}{\eta}\right]^{1/2} \tag{6.36}$$

6.8.3 *Probability of Error in PSK*

In BPSK, signal to represent symbol one is $S_1(t) = A\cos\omega_c t$ and signal to represent binary digit 0 is $S_0(t) = -A\cos\omega_c t$.

It is obvious, that, here the case of $S_0(t) = -S_1(t)$ implies. Therefore, Eqs. (3.44) and (3.47) are applicable here as follows.

$$\gamma^2{}_{\max} = \frac{8E_S}{\eta} \tag{6.37}$$

$$P_e|_{Min} = \frac{1}{2}erfc\left[\frac{E_S}{\eta}\right]^{1/2} \tag{6.38}$$

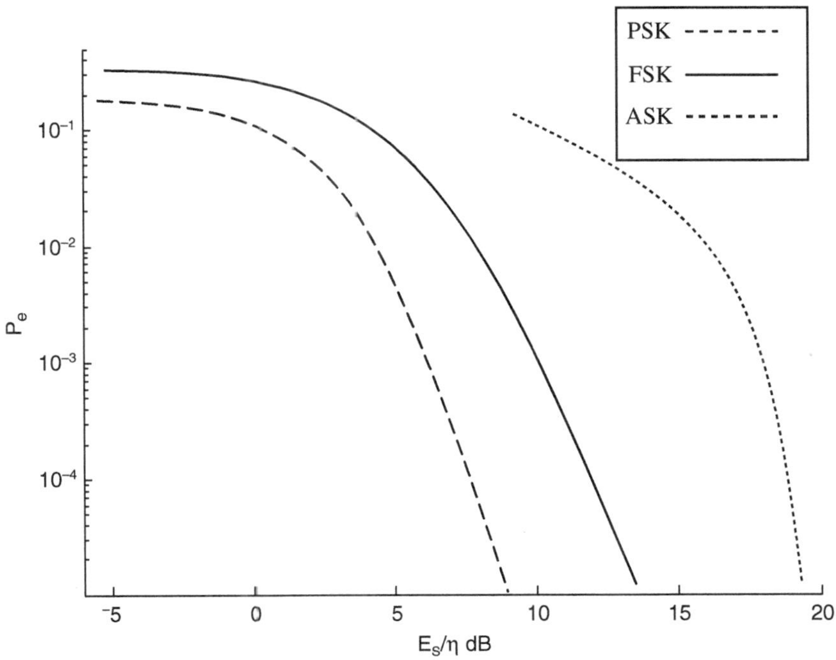

Fig. 6.30 Comparison of probability of error of ASK, PSK and FSK

6.9 MATLAB Programs

6.9.1 QPSK Waveform

```
% QPSK
%10110001;
% Ch6_QPSK
% Apurba Das
% 29/09/2009
% Output: Fig. 6.25
t=0:0.01:8;
a=[ones(1,200) zeros(1,601)];
T=1;
f=1/T;
% plot(t,a);
% figure;
sym=[ones(1,100) zeros(1,100) ones(1,200) zeros(1,300)
ones(1,101) ];
subplot(611);
plot(t,sym,'k');
```

```
axis off;
% title('Binary Data: 10110');
subplot(6,1,2);
s1=cos(2*pi*f*t-pi/4);
s1_1=s1.*a;
plot(t,s1_1);
subplot(6,1,3);
b=[zeros(1,200) ones(1,200) zeros(1,401)];
s2=cos(2*pi*f*t-3*pi/4);
s2_1=s2.*b;
plot(t,s2_1);

subplot(6,1,4);
b=[zeros(1,400) ones(1,200) zeros(1,201)];
s2=cos(2*pi*f*t-5*pi/4);
s3_1=s2.*b;
plot(t,s3_1);

subplot(6,1,5);
b=[zeros(1,600) ones(1,201)];
s2=cos(2*pi*f*t-7*pi/4);
s4_1=s2.*b;
plot(t,s4_1);

subplot(6,1,6);
ss=s1_1+s2_1+s3_1+s4_1;
plot(t,ss,'k');
title('\phi_{QPSK}(t)->');
```

6.9.2 MSK Waveform

```
% In CD: ch6_7.m
% MSK
% Output: Fig. 6.29
% Programed by Apurba Das (Sept, '09)

clc;
clear all;
close all;
t=0:.01:6.39; % 640 points

% Digital Signal
a=[ones(1,100) zeros(1,100) ones(1,200) zeros(1,101)];
```

```
T=6.39/8;
Tc=T/3;
f0=1/Tc;
% sI=cos((pi*t)/(2*T));
  sI=[ones(1,80) -ones(1,160) -ones(1,160) ones(1,160)
ones(1,80)].*cos((pi*t)/(2*T));
sQ=[ones(1,160) -ones(1,160) ones(1,160)
ones(1,160)].*sin((pi*t)/(2*T));
% sQ=sin((pi*t)/(2*T));
sI1=sI.*cos(2*pi*f0*t);
sQ1=sQ.*sin(2*pi*f0*t);
subplot(5,1,1);
plot(t,sI,'k');
subplot(5,1,2);
plot(t,sI1,'k');
subplot(5,1,3);
plot(t,sQ,'k');
subplot(5,1,4);
plot(t,sQ1,'k');
subplot(5,1,5);
plot(t,sQ1+sI1,'k');
```

References

1. Pasupathy, S., "Minimum Shift Keying: A Spectrally Efficient Modulation," *IEEE Communication Magazine*, July 1979
2. Chakrabarti, P., "Principles of Digital Communication", Dhanpat Rai and Co, 2008
3. Bedrosian, E., "Spectral Conservation by Efficient Channel Utilization", *Rand Corporation Report*, WN-9255-ARPA, Contract DAHC-15-73-C-0181. Santa Monica, California, October 1975
4. Skler, B., "Defining, Designing, and Evaluating Digital Communication Systems." *IEEE Communication Magazine*, Vol. 31, No. 11, 1993, pp. 92–101
5. Korn, I., "Principles of Coherent Communications", McGraw-Hill Book Co, 1966.
6. Lathi, B.P., "Modern Digital and Analog Communication Systems", Oxford University Press, 2005
7. Skler, B., "Digital Communications: Fundamentals and Applications (2nd Ed.)", Pearson Education, 2001

Chapter 7
Spread Spectrum Modulation

7.1 Introduction

In the technique of spread spectrum, the frequency spectrum of a data-signal (shaded in Fig. 7.1) is spread (non-shaded) using a code uncorrelated with that signal. As a result the bandwidth occupancy is much higher than required [4].

The codes used for spreading have low cross-correlation values and are unique to every user. This is the reason that a receiver which has knowledge about the code of the intended transmitter is capable of selecting the desired signal (Fig. 7.1).

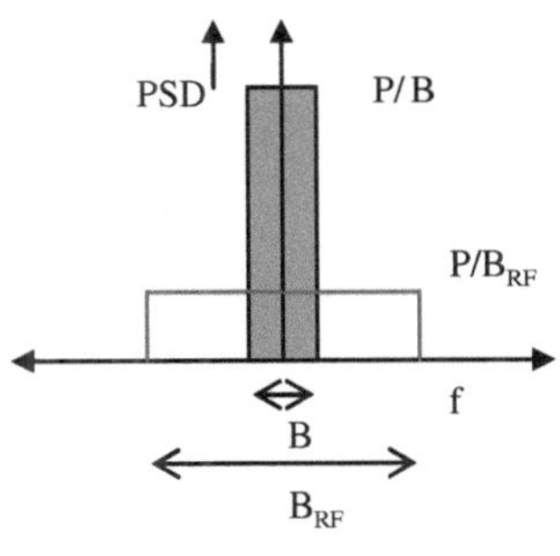

Fig. 7.1 PSD plot before and after spreading the signal

$$\text{Spreading factor} = \frac{B_{RF}}{B}$$

The points to be noted here are as follows:-

- √ The spread (i.e., modulated) signal can be de-spread by the intended receiver only.
- √ Spread spectrum technique is actually 2nd time modulation over an already modulated signal.

Say, an AM signal of power P_S has bandwidth 10 kHz. After spreading, it will have bandwidth 1 MHz.

Therefore, after spreading, the power is

A. Das, *Digital Communication*, Signals and Communication Technology,
DOI 10.1007/978-3-642-12743-4_7, © Springer-Verlag Berlin Heidelberg 2010

$$\mathrm{P_{sp}} = \frac{P_S}{\frac{10^6}{10^4}} = \frac{P_S}{100}$$

In dB,

$$\left.\frac{P_S}{100}\right\|_{\mathrm{dB}} = 10\log_{10} P_S - 10\log_{10} 100 = P_S\|_{\mathrm{dB}} - 20\ \mathrm{dB}$$

∴ Spread spectrum (modulated twice) provides noise 20 dB lower than the power of AM (modulated once).

A number of advantages are [2]:

- *Low power spectral density.* As the signal is spread over a large frequency-band, the Power Spectral Density is getting very small, so other communications systems do not suffer from this kind of communications. However the Gaussian Noise level is increasing.
- *Interference limited operation.* In all situations the whole frequency-spectrum is used.
- *Privacy due to unknown random codes.* The applied codes are - in principle – unknown to a hostile user. This means that it is hardly possible to detect the message of another user.
- *Applying spread spectrum implies the reduction of multi-path effects.*
- *Random access possibilities.* Users can start their transmission at any arbitrary time.
- Good anti-jam performance.

Because of the difficulty to jam or detect spread spectrum signals, the first applications were in the military field. However nowadays spread spectrum systems are gaining popularity also in commercial applications.

7.2 Processing Gain

The main parameter in spread spectrum systems is the processing gain: the ratio of transmission and information bandwidth:

$$\mathrm{G_p} = \frac{B_{RF}}{B} \tag{7.1}$$

which is basically the 'spreading factor'. The processing gain determines the number of users that can be allowed in a system, the amount of multi-path effect reduction, the difficulty to jam or detect a signal etc. For spread spectrum systems it is advantageous to have a processing gain as high as possible.

There exist different techniques to spread a signal:

1. Direct-Sequence (DS),
2. Frequency-Hopping (FH),
3. Time-Hopping (TH) and
4. Hybrid spread spectrum

We will concentrate on the two most popular techniques: Direct-Sequence and Frequency-Hopping, in this chapter.

7.3 Pseudo-Noise (PN) Sequence

PN sequence is the code used for spreading the signal of transmission. The frequency of the sequence is much higher with that of the transmitting signal. Actually the PN sequence is noise like, and periodic over one window. To understand the concept of pseudo-noise sequence let's consider one hypothetical experiment.

7.3.1 Concept: A Hypothetical Experiment

Let us consider a case where a faulty diode is generating random and high frequency noise with zero mean. For particular time duration, I have taken the noise in a form of a window. Now, the windows are concatenated side by side, one by one. These concatenated windows are given to one of the inputs of a zero-crossing detector (Fig. 7.2).

Thus, at the output, we will get a high frequency rectangular signal. Now, if we name the smallest chip duration (smallest high value time or smallest low value time) as logic 1 or logic 0 respectively, then we can get one chip sequence as 0011101| 0011101 | 001110. . .and so on (as the same windowed noise is concatenated, the sequence also made periodic).

From the sequence it's clear, that, the sequence is noise like random within a window, but periodic. That's why the sequence can be called as pseudo-random sequence (pseudo means 'not real').

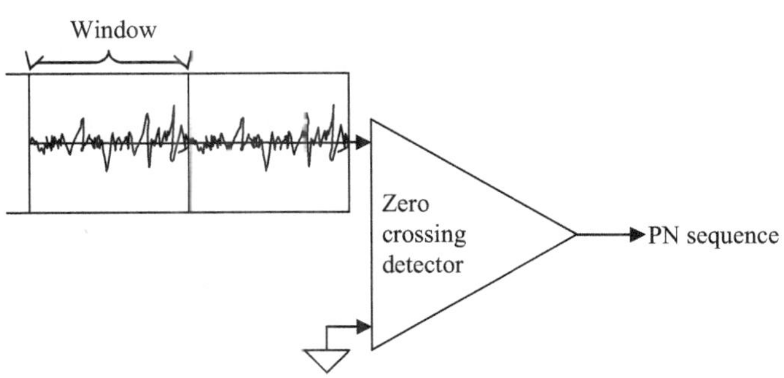

Fig. 7.2 Hypothetical experiment on concept of PN sequence

7.3.2 Generation of PN Sequence

In Fig. 7.3a, an m-bit length serial in parallel out shift register is shown [2, 5]. From all the flip-flops of the shift register, outputs are fed to one logic circuit with one switch to each. The output of the logic circuit is again fed to the input of the primary (left most) flip-flop. This arrangement is called as *feedback shift register* arrangement.

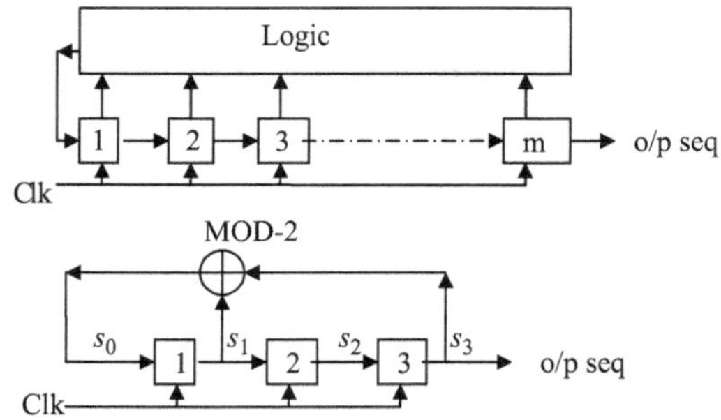

Fig. 7.3 (**a**) Feed back shift Register. (**b**) PN sequence generator

PN sequence can be easily generated by using the feedback register arrangement. Here I have considered 3 flip-flops with outputs s_1, s_2, s_3. Therefore the length of the shift register is m = 3. The output of the entire system is s_3. Its also can be said from Fig. 7.3b that, after a clock pulse is encountered, $s_2 = s_1, s_3 = s_2, s_1 = s_1 \oplus s_3$. The entire system state table is as follows, taking initial contents of the shift register '1111':

From Table 7.1, the shift register configuration shown in the Fig. 7.3b, is generating a sequence '1110100' followed by the same sequence again and again,

Table 7.1 State table for PN sequence generation

Clock pulse no.	Intermediate states			Output
	S_1	S_2	S_3	
1	1	1	1	1
2	0	1	1	1
3	1	0	1	1
4	0	1	0	0
5	0	0	1	1
6	1	0	0	0
7	1	1	0	0
8	1	1	1	1

periodically. It is observed that after 8th clock pulse we are getting the same initialized flip flop content '111'.
Note: MOD-2 addition:

MOD-N means periodic with periodicity N. Therefore MOD-2 operation generates a window of sequence of length 2. To represent 2 patterns we need only one bit (0 and 1). Let's consider MOD-2 addition (i.e., one bit number + one bit number = one bit number).

$$0 + 0 = 0$$
$$0 + 1 = 1$$
$$1 + 0 = 1$$
$$1 + 1 = 10$$

The last result (1+1) is generating two bit result. The MSB is called as CARRY and the LSB is the SUM. To follow the procedural principle of MOD-2 addition, we must have only 1 bit. Therefore, by ignoring the CARRY bit, the result of (1+1) operation is made equal to 0. The mod-2 addition is symbolized by$\oplus$. The truth table is shown in Table 7.2.

Table 7.2 Table of MOD-2 operation is following the XOR gate truth table

A	B	A⊕B
0	0	**0**
0	1	**1**
1	0	**1**
1	1	**0**

From the Table 7.2, it is very clear that, MOD-2 operation is following the XOR gate truth table. Therefore the MOD-2 operation is realized practically by XOR gate.

7.3.3 Properties of PN Sequence

1. In each period, Number of 1s = Number of zeroes +1;
2. If the number of flip-flop is m, the sequence length N must be, $N = (2^m - 1)$
3. At any instant, the square of the PN sequence c(t) is 1. $c^2(t) = 1$.
4. Autocorrelation

$$R_c(\tau) = \frac{1}{T_c} \int_{-\frac{T_c}{2}}^{\frac{T_c}{2}} c(t)c(t-\tau)dt,$$

where T_c is the chip duration of the PN sequence. $T_c \ll T_b$ (bit duration). The derivation of auto correlation can be derived from the table as follows:

Table 7.3 Autocorrelation function table of PN sequence

Shift	Original/shifted	$\sum$	$R_c(\tau)=\sum/N$
0	1110100 1110100	1+1+1+1+1+1+1=7	7/7=1
1	1110100 0111010	−1+1+1−1−1−1+1=−1	−1/7= −0.14
2	1110100 0011101	−1−1+1−1+1−1+1=−1	−1/7=−0.14
3	1110100 1001110	+1−1−1−1+1−1+1=−1	−1/7=−0.14
4	1110100 0100111	−1+1−1+1+1−1−1=−1	−1/7=−0.14
5	1110100 1010011	+1−1+1+1−1−1−1=−1	−1/7=−0.14
6	1110100 1101001	+1+1−1−1−1+1−1=−1	−1/7=−0.14
7	1110100 1110111	+1+1+1+1+1+1=7	7/7=1
8	1110100 0111010	−1+1+1−1−1−1+1=−1	−1/7=−0.14

The graph obtained from the Table 7.3, is as follows (Fig. 7.4).

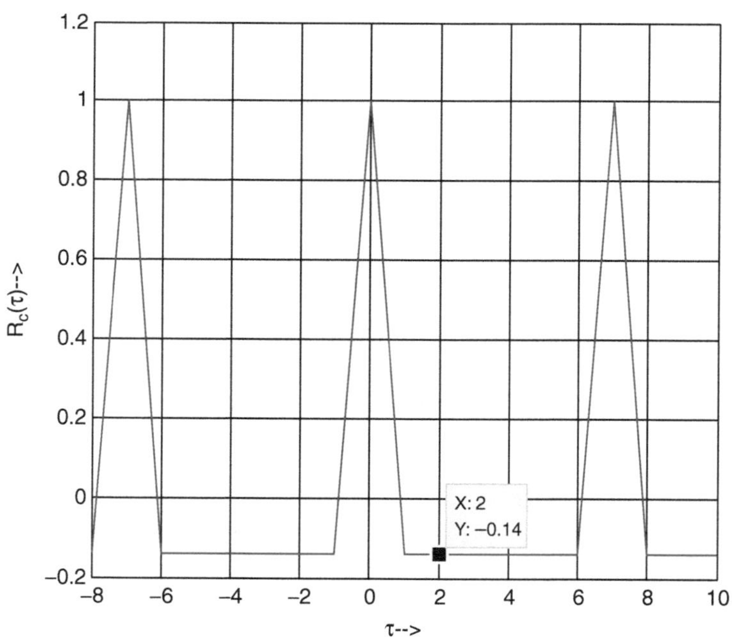

Fig. 7.4 Autocorrelation of PN sequence

7.4 Direct Sequence Spread Spectrum (DSSS)

7.4.1 Concept

Direct Sequence is the best known Spread Spectrum Technique [4]. Here, the data signal (with bit duration T_B) is multiplied by a Pseudo Random Noise Code (PN code) with chip duration T_C, where $T_C \ll T_B$.

A PN code is a sequence of chips valued −1 and 1 (polar) or 0 and 1 (non-polar) and has noise-like properties. This results in low cross-correlation values among the codes and the difficulty to jam or detect a data message.

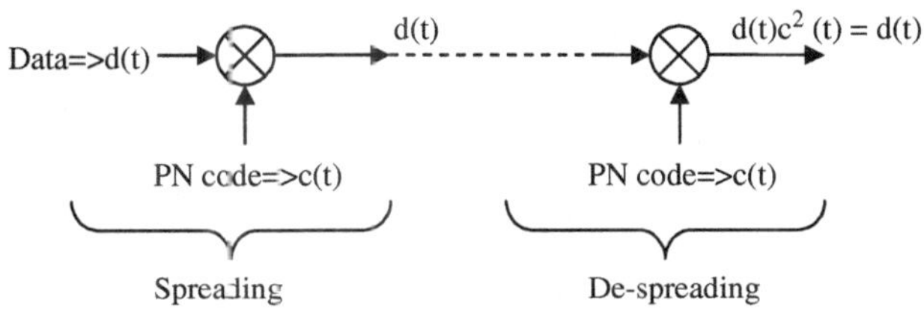

Fig. 7.5 Direct sequence spread spectrum

As shown in the Fig. 7.5, the data d(t) is multiplied by the high frequency chip sequence (PN code) c(t). It results in bandwidth spreading.

$$\text{As,}\ c^2(t) = 1,$$

if, the spread sequence is again multiplied by c(t), the output of the receiver would be $d(t)c^2(t) = d(t)$. The chips are actually chopping the bit sequence s shown in the Fig. 7.6.

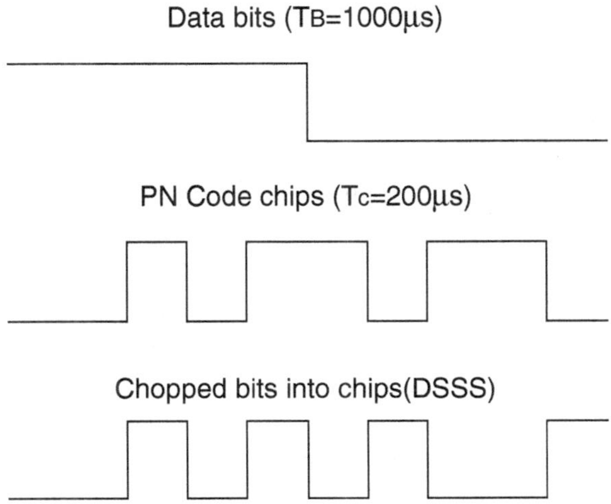

Fig. 7.6 Direct-sequence spreading by bit chopping

As the bandwidth of the data signal is now multiplied by a factor G_P . The power contents however stays the same, with the result that the power spectral density lowers.

Here, the spreading is done by multiplying the digital data with PN chip sequence. This is very much analogous to Amplitude Modulation technique as (1) the modulation is done by simple multiplication and (2) the amplitude of the data is being modified/modulated, both of which are characteristics of Amplitude Modulation (AM).

For this reason it is easy to introduce a large processing-gain in Direct-Sequence systems.

In the receiver, the received signal is multiplied again by the same (synchronized) PN code. Since the code existed of +1s and −1s, this operation completely removes the code from the signal and the original data-signal is left. Another observation is that the de-spread operation is the same as the spread operation. The consequence is that a possible jamming-signal in the radio channel will be spread before data-detection is performed. So jamming effect is reduced (see Fig. 7.7).

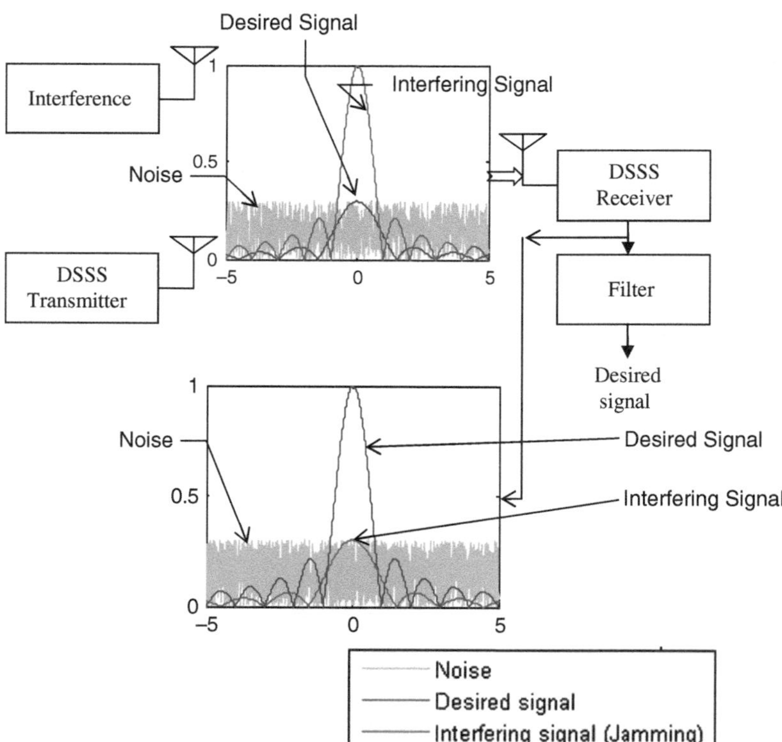

Fig. 7.7 DS-concept, before and after despreading

The main problem with applying Direct Sequence spreading is the so-called Near–Far effect which is illustrated in Fig. 7.8. This effect is present when an interfering transmitter is much closer to the receiver than the intended transmitter. Although the cross-correlation between codes A and B is low, the correlation between the received signal from the interfering transmitter and code A can be higher than the correlation between the received signal from the intended transmitter and code A. The result is that proper data detection is not possible.

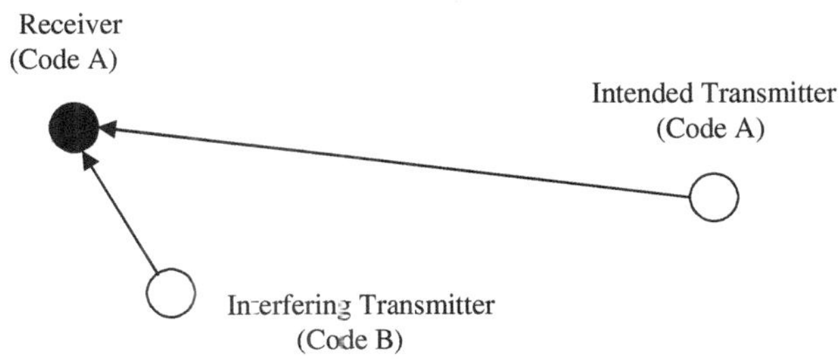

Fig. 7.8 Illustration of Near–far effect

7.4.2 DSSS with Coherent BPSK

As we know from previous chapters, Binary Phase Shift Keying is a very important digital modulation technique in practice. Here one BPSK modulate signal is treated as the information s(t) and multiplied by the chip sequence (PN code) c(t). As DSSS is also a modulation technique, we can infer, this technique of spread spectrum modulation is actually 2nd time modulation over an already modulated wave.

From the Fig. 7.9, we get the BPSK signal s(t). The signal transmitted through the channel is s(t)c(t). Additive white Gaussian Noise (AWGN) n(t) is added in the channel. Therefore the received signal is

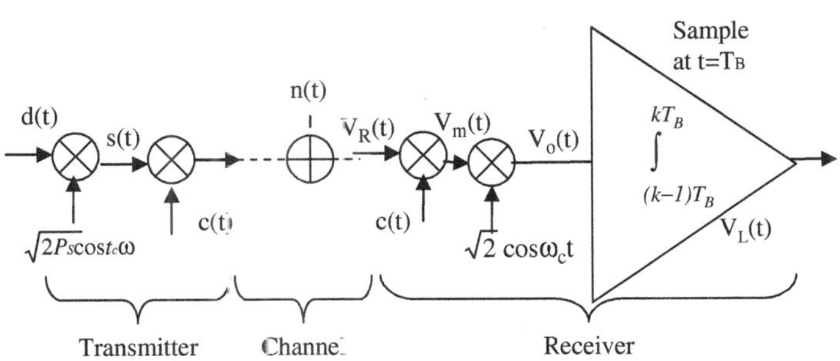

Fig. 7.9 BPSK incorporating DSSS

$$\begin{aligned} v_R(t) &= s(t)c(t) + n(t) \\ \Rightarrow v_R(t) &= \sqrt{2P_S}d(t)c(t)\cos\,\omega_c t + n(t) \end{aligned} \tag{7.2}$$

After multiplying the received signal by PN chip code c(t) we get,

$$\begin{aligned} v_m(t) &= \sqrt{2P_S}d(t)c^2(t)\cos\,\omega_c t + n(t)c(t) \\ \Rightarrow v_m(t) &= \sqrt{2P_S}d(t)\cos\,\omega_c t + n(t)c(t) \end{aligned} \tag{7.3}$$

The 2nd term of the Eq. (7.3) indicates the noise is chopped, i.e. (1) noise is spread whereas wanted data is de-spread and hence (2) power spectral density lowers as shown in the Fig. 7.7.

Next, $v_m(t)$ is again passed through another multiplier, integrator, i.e. BPSK detector. The probed signals are as follows.

$$v_o(t) = \sqrt{P_S}(2\cos^2\,\omega_c t)d(t) + n(t)c(t)\cos\,\omega_c t \tag{7.4}$$

The signal $v_0(t)$ is then sampled at $t=kT_B$ to get the final output $v_L(t)$.

$$\begin{aligned} v_L(t) &= \int_0^{T_B} v_o(t)dt; k = 1 \\ \Rightarrow v_L(t) &= \int_0^{T_B} \left[\sqrt{P_S}(2\cos^2\,\omega_c t)d(t) + n(t)c(t)\cos\,\omega_c t\right]dt = d(t) \end{aligned} \tag{7.5}$$

By this way the data signal d(t) has been correctly detected at the end output.

7.4.3 Probability of Error Calculation

From the discussions of Chaps 3 and 6, it has been shown that, the probability of error of BPSK for matched filter reception without employing spread spectrum is

$$P_e = \frac{1}{2}erfc\left[\frac{E_S}{\eta}\right]^{1/2} \tag{7.6}$$

where, E_S is the symbol energy, $\eta/2$ is the two sided power spectral density (variance) of the noise.

To consider the case of direct sequence spread spectrum incorporated with BPSK (Eq. 7.2), within bit interval $0 \le t \le T_B$, the transmitted signal is

$$s_t(t) = d(t)g(t)c(t)\cos 2\pi f_c t; 0 \le t \le T_B \tag{7.7}$$

where, d(t) is the information symbol and the pulse g(t) is defined from the Eq. (7.2) as

$$g(t) = \begin{cases} \sqrt{2P_S} = \sqrt{\dfrac{2E_S}{T_B}}; 0 \le t \le T_B \\ 0; \qquad\qquad \text{otherwise} \end{cases} \tag{7.8}$$

and c(t) is the PN code generated, in the bit interval $0 \le t \le T_B$ expressed as

$$c(t) = \sum_{n=0}^{L_C-1} c_n p\,(t - nT_C) \tag{7.9}$$

Here,

L_c=number of chips per bit
$\{c_n\}$=PN sequence

The code chip sequence is uncorrelated (white), i.e.,

$$E\,(c_n c_m) = 0; \quad for \quad n \neq m \tag{7.10}$$

and each chip is +1 or −1 with probability 0.5.

$$\therefore \quad E\,(c_n) = 0, \quad \text{and} \quad E\left(c_n^2\right) = 1 \tag{7.11}$$

When the signal is interfered by an additive signal n(t), the received signal is

$$r(t) = d(t)g(t)c(t)\cos 2\pi f_c t + n(t) \tag{7.12}$$

Since the received signal is the output of a bandpass signal, n(t) must have only the band pass components, i.e.,

$$n(t) = n_c(t)\cos 2\pi f_c t - n_s(t)\sin 2\pi f_c t \tag{7.13}$$

Equation (7.13) is the expression of bandpass random noise. $n_c(t)$ and $n_s(t)$ are the two quadrature components.

After this, the received signal is multiplied by c(t) and goes through an integrator, as shown in Fig. 7.9. Taking out only the received noise part from Eq. (7.4), sampled at t=T_B,

$$
\begin{aligned}
y_n(T_B) &= \int_0^{T_B} c(t)n(t)g(t)\cos(2\pi f_c t)\,dt \\
&= \sum_{n=0}^{L_C-1} c_n \int_0^{T_B} p(t-nT_c)n(t)g(t)\cos(2\pi f_c t)\,dt \\
&= \sqrt{\frac{2E_S}{T_B}} \sum_{n=0}^{L_C-1} c_n s_n
\end{aligned} \tag{7.14}
$$

where,

$$
s_n = \int_{nT_c}^{(n+1)T_c} n(t)\cos(2\pi f_c t)\,dt \tag{7.15}
$$

The probability of error directly depends upon the statistical parameters of the interfering signal. The mean of the noise is

$$
E\left[y_n(T_B)\right] = 0 \tag{7.16}
$$

And the variance is

$$
E\left[y_n^2(T_B)\right] = \frac{2E_S}{T_B} \sum_{n=0}^{L_C-1} \sum_{m=0}^{L_C-1} E(c_n c_m) E(s_n s_m)
$$

As $E(c_n c_m) = \delta_{nm}$, as the property of PN sequence suggests, therefore,

$$
E\left[{y_n}^2(T_B)\right] = \frac{2E_S}{T_B} \sum_{n=0}^{L_C-1} E\left({s_n}^2\right) \tag{7.17}
$$

Now, considering a single tone interference, i.e.,

$$
n(t) = \sqrt{2P_N}\cos 2\pi f_c t \tag{7.18}
$$

where, P_N is the average noise power. Substituting Eq. (7.18) into Eq. (7.15),

$$
s_n = \sqrt{2P_N} \int_{nT_c}^{(n+1)T_c} \cos^2(2\pi f_c t)\,dt = \frac{\sqrt{2P_N}}{2} \int_{nT_c}^{(n+1)T_c} dt = \frac{T_c\sqrt{P_N}}{\sqrt{2}} \tag{7.19}
$$

Here also, mean of noise

$$E(s_n) = E\left(\frac{T_C\sqrt{P_N}}{\sqrt{2}}\right) = 0 \tag{7.20}$$

The mean square value of noise is

$$E\left({s_n}^2\right) = \frac{{T_C}^2 P_N}{2} \tag{7.21}$$

From Eqs. (7.21) and (7.17), we get the noise power/variance is

$$E\left[{y_n}^2\left(T_B\right)\right] = \frac{E_S P_N T_c}{2} \tag{7.22}$$

The ratio of signal power and noise power is the signal-to-noise ratio at the detector expressed as

$$\text{SNR} = \frac{E_S^2}{E_S P_N T_c/2} = \frac{2E_S}{P_N T_c} \tag{7.23}$$

If the noise variance $\eta/2$of Eq. (7.6) is same as $\frac{E_S P_N T_c}{2}$, then the probability of error of the system is

$$P_e = \frac{1}{2} erfc\left(\frac{E_S}{E_s P_N T_c}\right)^{1/2} = \frac{1}{2} erfc\left(\frac{f_c}{P_N}\right)^{1/2} \tag{7.24}$$

The expression of probability of error signifies, as chip-frequency is very high, the probability of error must be very low, according to the nature of the *erfc* graph shown in the Chap. 3.

7.5 Frequency-Hopped Spread Spectrum

7.5.1 Concept

Remembering FM in analog communication or FSK in digital communication, the concept of frequency hopped spread spectrum (*FHSS*) can be understood. In FM, the instantaneous carrier frequency is changed according to the baseband signal, whereas in FHSS, the input carrier frequency to the modulator itself varies within a fixed bandwidth. With respect to time, the frequency assigned for modulation is changed with a central frequency but with a fixed bandwidth. As the frequencies are changed from one to another and the allocation of frequency is pseudorandom, i.e., not in order (dependent upon a PN sequence), the term '*hopping*' comes to represent the allocation of frequency with respect to time [6]. By this carrier frequency hopping, obviously the bandwidth of the signal increased.

The process of frequency hopping is illustrated in Fig. 7.10. A disadvantage of Frequency-Hopping over Direct-Sequence is that obtaining a high processing-gain is hard. The faster the 'hopping-rate' is, the higher the processing gain.

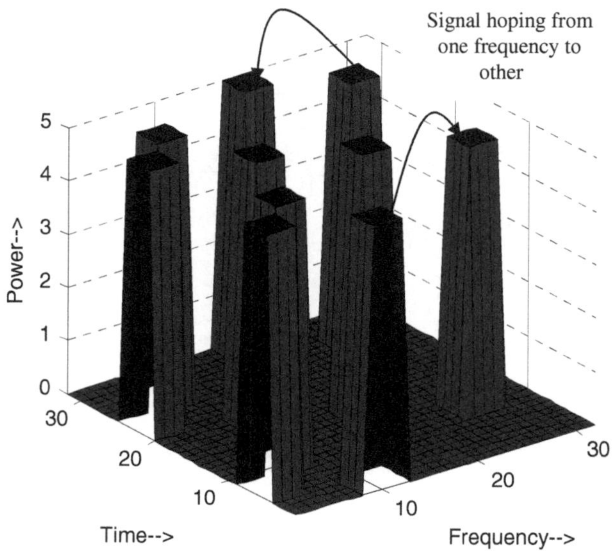

Fig. 7.10 Illustration of the frequency hopping concept

On the other hand, Frequency-Hopping is less affected by the Near–Far effect than Direct-Sequence. Frequency-Hopping sequences have only a limited number of 'hits' with each other. This means that if a near-interferer is present, only a number of 'frequency-hops' will be blocked in stead of the whole signal. From the 'hops' that are not blocked it should be possible to recover the original data-message.

Figure 7.11 shows the general scheme of FHSS modulation. A PN sequence generator creates a k-bit pattern for every hopping period, T_h. The frequency table uses the pattern to find the frequency to be used for a particular hopping period and passes it to the frequency synthesizer. The frequency synthesizer creates a carrier signal of that frequency, and the source signal modulates the carrier signal.

Let's consider a case (for theoretical illustration) where the PN code generator creates three bit patterns (8 different patterns). The 8 patterns generated by PN code generator are mapped to eight different frequencies in the frequency table (Fig. 7.12).

The pattern for this station is 101, 111, 001, 000, 010, 011, and 100. Note that the pattern is pseudorandom it is repeated after eight hoppings. This means that at hopping period 1, the pattern is 101. The frequency selected is 700 kHz; the source signal modulates this carrier frequency [3]. The second k-bit pattern selected

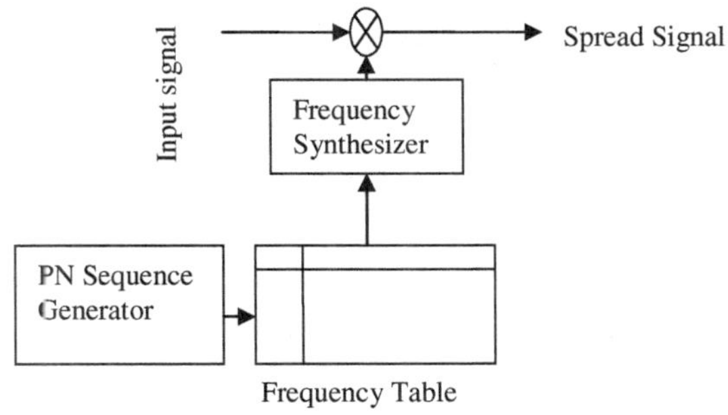

Fig. 7.11 FHSS modulator

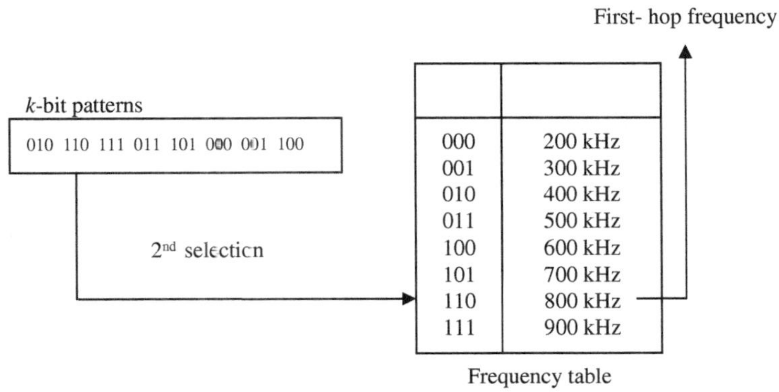

Fig. 7.12 Frequency selection in FHSS

is 111, which selects the 900-kHz carrier; the eighth pattern is 100, the frequency is 600 kHz. After eight hoppings, the pattern repeats, starting from 101 again.[1]

7.5.2 FHSS with FSK

In short, the technique of integrated modulation by FHSS and FSK is discussed with proper block diagram representation of transmitter and receiver. As we know from the chapter of digital modulation, FSK is very much like FM. The carrier frequency used for frequency modulation is here dynamically determined by the PN sequence generator as discussed in the previous section.

We know that, the general FSK (before applying FHSS) can be expressed as

$$s_{\mathrm{FSK}}(t) = \sqrt{2P_S}\cos\{\omega_c t + d(t) \times \Omega t + \theta\} \tag{7.25}$$

[1]Coherent detection of FHSS is possible only within each hop. Therefore, non-coherent detection is employed.

After the second modulation by FHSS of the already modulated (FSK) signal, the signal becomes:

$$s_{FSK\_FHSS}(t) = \sqrt{2P_S}\cos\{\omega_i t + d(t) \times \Omega t + \theta\} \tag{7.26}$$

where, ω_i =instantaneous carrier frequency. The resultant signal is therefore responsible for *instantaneous transmission bandwidth.*
As shown in Fig. 7.13, here also, successive k bit segments of PN sequence drive the frequency synthesizer which enables the carrier frequency to hop over 2^k distinct values.

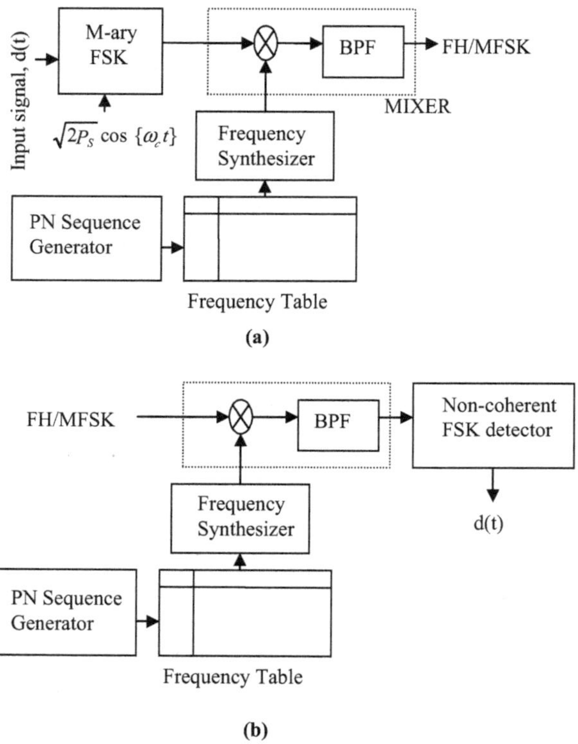

Fig. 7.13 Frequency hop spread M-ary FSK. (**a**) Transmitter, (**b**) Receiver

Each hop is obviously of conventional M-FSK. Therefore, through complete range of 2^k hops, the MFSK signal occupies much higher bandwidth.

We know from the previous discussion that, coherent detection is possible only within each hop, as the pilot carrier frequency is being changed according to the generated PN sequence. Therefore, non-coherent detection is employed.

7.5.3 Rate of Hopping: Fast and Slow

In FHSS, the available channel bandwidth W is sub-divided into a large number of non-overlapping (orthogonal) frequency slots. In any signaling interval, the transmitted signal occupies one or more of the available frequency slots. The selection of frequency slots in each signaling interval is made pseudorandomly according to the output from the PN generator. This pseudo non-deterministic change or jump of frequency is *frequency hopping* and the rate of frequency slot change is the hopping rate (R_h). As for an example in the case of binary FSK ($M = 2$), the modulator selects 2 frequencies corresponding to transmission of ‘0’ or a ‘1’ symbol [6].

The term *slow frequency hopping (SFH)* means there are several modulation symbols per hop. Similarly, *fast frequency hopping (FFH)* means there are several frequency hops per modulation symbol. For SFH, the shortest uninterrupted waveform of the system is that of the data symbol; whereas, for FFH, the shortest uninterrupted waveform of the system is that of the hop.

As an example of FFH, the aforementioned binary FSK system is taken. According to the Fig. 7.14a, there are 4 chips transmitted per bit. Here, for FFH, the chip duration itself is the hop duration. Fig. 7.14b illustrates an example of SFH binary FSK system. In this case, there are 3 bits transmitted during the time duration of single hop. Here, the hop duration is three times the chip duration. If we employ a 4-ary FSK ($M = 4$), then the number of bits/MSK symbol is $k = 2$. If 3 PN sequence chips are accommodated into single hop, then frequency allocation diagram shown in Fig. 7.15 can be obtained. In this case, the symbol duration (duration of 4 bits) and the hop duration will be identical.

7.6 Application of Spread Spectrum

7.6.1 GPS (Global Positioning System)

The global positioning system (GPS) uses direct sequence spread spectrum and operates on the principle of one way *ranging*. It is divided into three segments namely, space, control and user segment. The selected satellite orbital constellation contains 24 satellites in 6 orbital planes. The six orbital planes are 60° apart from each other containing 4 satellites each. The satellites have a period of 12 h sidereal time and a semi-major axis of 26561.75 km. One sidereal day means 23 h 56 min 4.09s which is the interval between two successive transits of the vernal equinox and is defined as the mean sidereal day. The four satellites are unevenly placed in each plane so as to minimize the effect of satellite outage. The eccentricity of the orbit is nearly zero and the inclination of the orbital plane is 55°. This satellite constellation allows at least four satellites to be visible to a user anywhere anytime on the earth. Table 7.4 lists the orbital parameters of the GPS constellation. The satellite visibility is also depicted in Fig. 7.16.

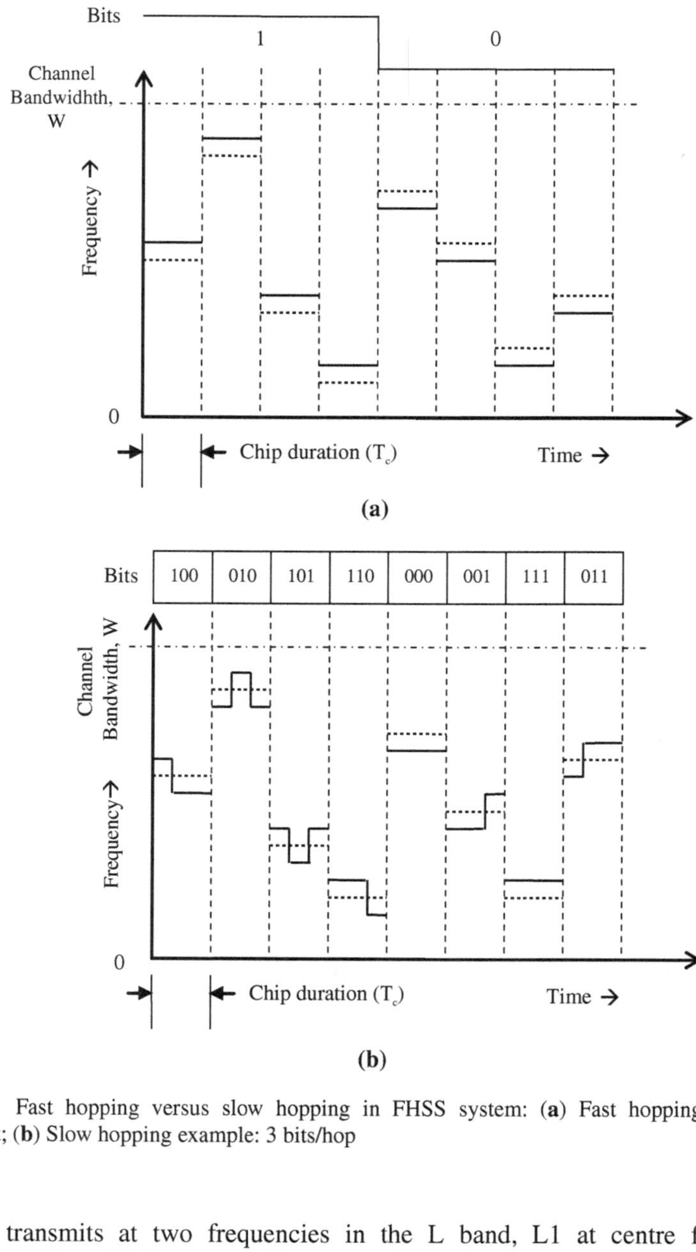

Fig. 7.14 Fast hopping versus slow hopping in FHSS system: (**a**) Fast hopping example: 4 hops/bit; (**b**) Slow hopping example: 3 bits/hop

GPS transmits at two frequencies in the L band, L1 at centre frequency 1,575.42 MHz and L2 at centre frequency 1,227.6 MHz (each being a multiple of 10.23 MHz). The L1 signal is bi-phase modulated. The spreading code with I channel is a Gold code of length 1,023 chips. The rate is 1.023 MHz and code period is 1 ms. Each satellite uses pre-assigned gold code. The code is called as course/acquisition or simply C/A code. The Q channel spreading code is a long non-linear code frequency 10.23 MHz. The code is called as precision or P code.

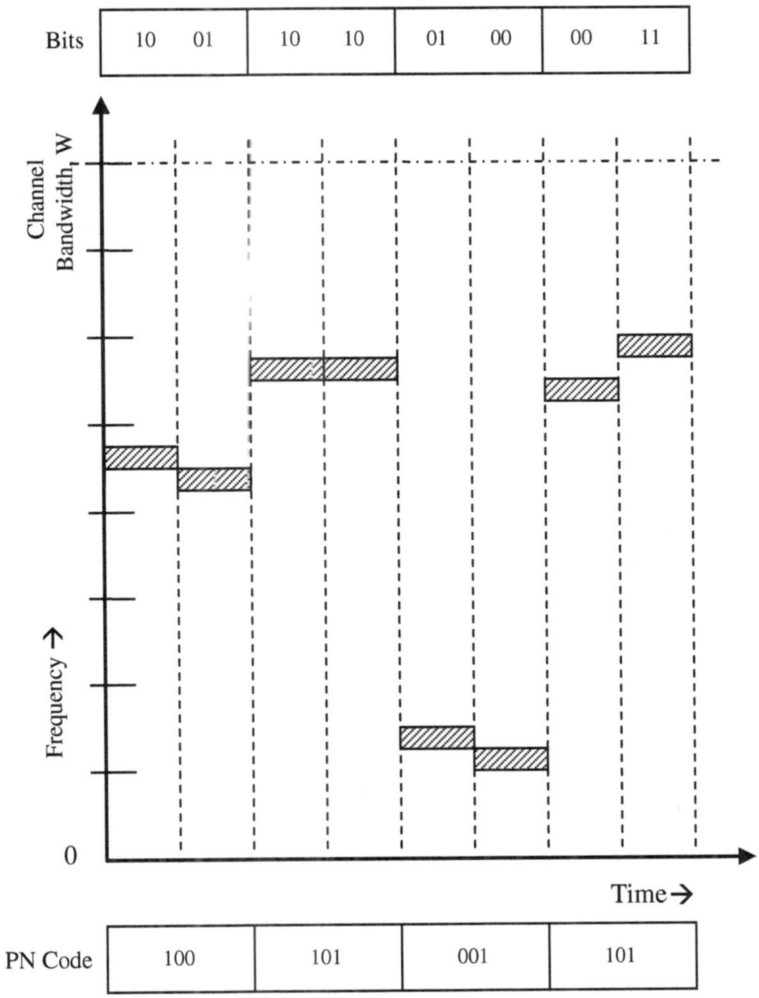

Fig. 7.15 Frequency variation over a complete period of the PN sequence

It is transmitted at a rate of 50 bps and repeated cyclically. The message contains information about the positions of the satellites, data to correct errors in propagation time, satellite clock bias, etc. a method of producing L1 signal is depicted in the Fig. 7.17.

The control segment consists of 5 monitor stations (Hawaii, Colorado Springs, Ascension Islang, Diego Gracia and Kwajalan Island), four ground stations and operational control centre. All the monitor stations except Hawaii contain ground antenna upload stations. The clocks of the satellites are synchronized by a ground station of California. It also supplies the positions of the satellites obtained from

Table 7.4. Orbital parameters of the GPS constellation

No.	Satellite identification	Semi-major axis	Inclination, θ	Right ascension of ascending node	Mean anomaly	Phase difference	Longitude of ascending node
1.	A3	26,561.75	55.0	272.847	11.676	103.55	179.63
2.	A4	26,561.75	55.0	272.847	41.806	31.13	14.69
3.	A2	26,561.75	55.0	272.847	161.786	119.98	74.68
4.	A1	26,561.75	55.0	272.847	268.126	106.34	127.85
5.	B1	26,561.75	55.0	272.847	80.956	130.98	94.27
6.	B2	26,561.75	55.0	272.847	173.336	92.38	140.46
7.	B4	26,561.75	55.0	272.847	204.376	31.04	155.98
8.	B3	26,561.75	55.0	272.847	309.976	105.06	28.78
9.	C1	26,561.75	55.0	272.847	111.876	100.08	169.73
10.	C4	26,561.75	55.0	272.847	241.556	129.68	54.57
11.	C3	26,561.75	55.0	272.847	339.666	98.11	103.62
12.	C2	26,561.75	55.0	272.847	11.796	32.13	119.69
13.	D1	26,561.75	55.0	272.847	135.226	100.07	61.40
14.	D4	26,561.75	55.0	272.847	167.356	32.13	77.47
15.	D2	26,561.75	55.0	272.847	265.446	98.09	127.51
16.	D3	26,561.75	55.0	272.847	35.156	129.71	11.37
17.	E1	26,561.75	55.0	272.847	197.046	130.98	152.31
18.	E2	26,561.75	55.0	272.847	302.596	105.55	25.09
19.	E4	26,561.75	55.0	272.847	333.686	31.09	40.63
20.	E3	26,561.75	55.0	272.847	66.066	92.38	6.82
21.	F1	26,561.75	55.0	272.847	238.886	103.54	53.23
22.	F2	26,561.75	55.0	272.847	345.226	106.34	106.40
23.	F3	26,561.75	55.0	272.847	105.206	119.98	166.39
24.	F4	26,561.75	55.0	272.847	135.346	30.00	1.46

tracker system. Therefore, each satellite knows the position of all GPS satellites at any instant.

The user segment consists of GPS receiver. The receiver consists of a pre-amplifier, antenna, down converter, A/D converter, IF amplifier, navigation processing unit.

Fig. 7.16 GPS constellation in motion with the rotation of the earth

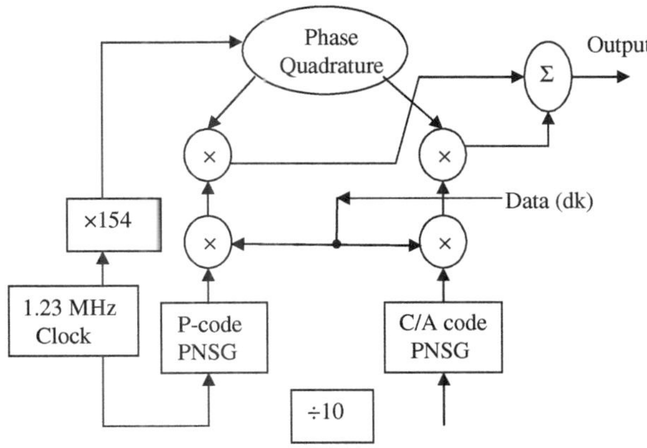

Fig. 7.17 Arrangement of producing L1 signal

Circular or rectangular microstrip is the common choice of antenna. The received GPS signal is very weak (of the order of −130 dB). But the reception is possible as the characteristics is spread spectrum, even if the signal is below the noise level.

The GPS receiver has a single channel and demodulates the signals from each of the four satellites one by one. A receiver determines the range of the satellite using the C/A code or P code.

7.7 CDMA (Code Division Multiple Access)

7.7.1 Orthogonal Chip Sequence

As the name signifies, code division multiplexing is a process of multiplexing where, a number of input channels are multiplexed on the basis of codes, i.e., codes are used as carriers of modulation in the transmitter end [1]. Here, 'code' signifies pseudo-random sequences, chosen properly according to some constrains of orthogonality. In case of spread spectrum modulation, pseudorandom chip sequences are used for bandwidth spreading. Therefore, the frequency of chip sequence should be much higher than the frequency of data stream; i.e., bit duration (T_b)<chip duration (T_c). Here, in case of CDM, direct sequence spread spectrum modulation is used in the transmitter side. But, the constraint is, the chip sequences should be properly defined and orthogonal, so that, no one channel modulated by code sequence 1, can be interfered by other channel modulated by code sequence 2. To satisfy the constraint, the sequences need to follow two distinct properties as follows:

(i) Cross-correlation Sum of Product is Zero:

If we consider the format of chip and bit sequences as bipolar NRZ, logic 1 is represented by +1V and logic zero is represented by −1. Now, if we choose any two sequences of 4 digits, arbitrarily, result of addition of bit by bit product would be 0.

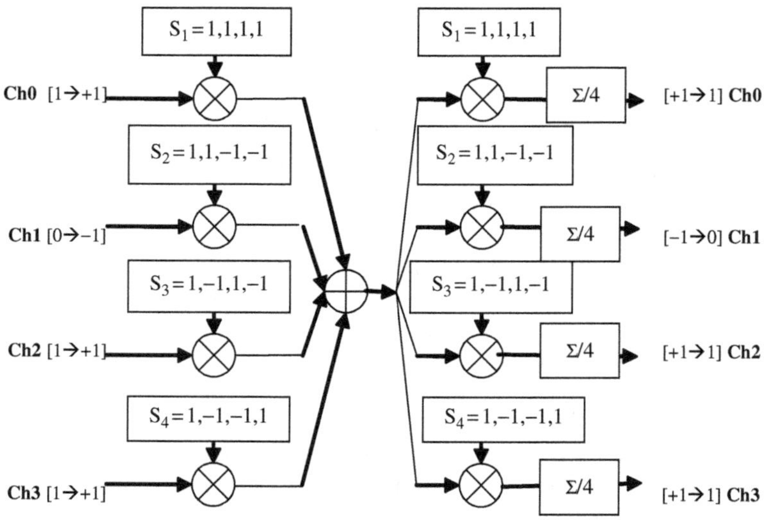

Fig. 7.18 DSSS based CDMA

As for an example (Fig. 7.18), take any two sequences 1010 and 1100.

$$SOP_{cross\_bit} = \frac{1}{4}\sum (1,-1,1,-1)\,.^{*}\,(1,1,-1,-1),$$

where, .* is the bit by bit product (analog multiplication) operator

$$\therefore SOP_{cross\_Bit} = \frac{1}{4}\left\{(1^{*}1) + (-1^{*}1) + (1^{*}-1) + (-1^{*}-1)\right\}$$
$$= \frac{1}{4}\{1-1-1+1\} = 0$$

We can test the claim by choosing another pair, say 1001, 1100

$$SOP_{cross\_Bit} = \frac{1}{4}\sum (1,-1,-1,1)\,.^{*}\,(1,1,-1,-1)$$
$$= \frac{1}{4}\left\{(1^{*}1) + (-1^{*}1) + (-1^{*}-1) + (1^{*}-1)\right\}$$
$$= \frac{1}{4}\{1-1+1-1\} = 0$$

(ii) Auto-correlation Sum of Product is 'One':

Now, if we choose any sequence of 4 digits (N = 4), result of addition of bit by bit product would be.

As for an example, take any two sequences 1010.

$$SOP_{auto\_bit} = \frac{1}{N}\sum (1,-1,1,-1)\,.^{*}\,(1,-1,1,-1),$$

where, .* is the bit by bit product (analog multiplication) operator

$$\therefore SOP_{auto\_Bit} = \frac{1}{4}\left\{(1^{*}1)+(-1^{*}-1)+(1^{*}1)+(-1^{*}-1)\right\}$$
$$= \frac{1+1+1+1}{4} = 1$$

The chip sequences which follow these two relative constraints are called as orthogonal chip sequences. In our discussion, we have chosen 4 orthogonal chip sequences of 4 digits as, 1111, 1100, 1010, 1001.

7.7.2 Gold Sequence

From the discussion of previous sub-section it is easily understandable that, the number of orthogonal chip sequence increases, as the length of the sequence increases. But, most of the sequences are not suitable for CDMA use. Prof. Kasami and Gold have obtained PN-sequences with better periodic cross-correlation properties. They are derived from the all possible sequences of a particular length and are known as *Gold sequences*. [7, 8]

Gold sequence offer a large number of sequences set with good cross-correlation properties between each pairs. Such sequences are generated by modulo-2 addition of two sequences driven by the same clock as shown in the Fig. 7.19. The most important factor of designing Gold sequence is only special pairs of allowed sequences can give the desired correlation properties.

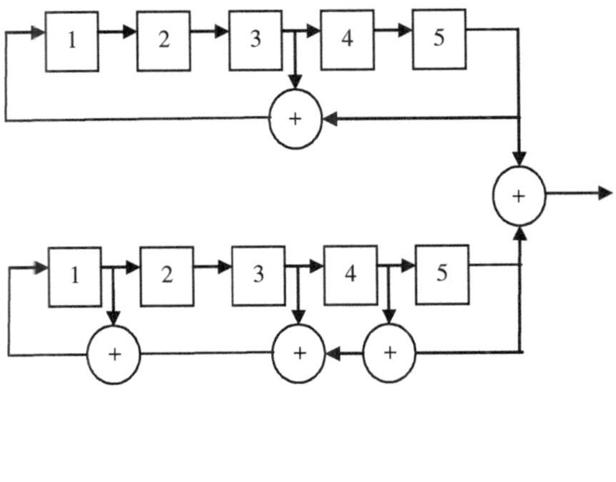

Fig. 7.19 Gold sequence generation

7.7.3 Principle of Operation

7.7.3.1 MUX

According to the figure above (Fig. 7.18), 4 channels are multiplexed into one trunk. Bits sent from Ch0, Ch1, Ch2 and Ch3 are 1, 0, 1, and 1 respectively [1]. These 4 symbols are modulated (DSSS) using seq0 (1111), seq1 (1100), seq2 (1010) and seq3 (1001) respectively, all are in BNRZ format. Therefore, the results of 4 modulators would be

$$\text{Ch0\_MOD} \rightarrow (+1, +1, +1, +1)^{*}1 = (+1, +1, +1, +1)$$
$$\text{Ch0\_MOD} \rightarrow (+1, +1, -1, -1)^{*}-1 = (-1, -1, +1, +1)$$
$$\text{Ch0\_MOD} \rightarrow (+1, -1, +1, -1)^{*}1 = (+1, -1, +1, -1)$$
$$\text{Ch0\_MOD} \rightarrow (+1, -1, -1, +1)^{*}1 = (+1, -1, -1, +1)$$

Adding all the above sequences bit by bit, we get, (2, −2, 2, 2). This is indeed the multiplexed sequence. We can say, this sequence is the multiplexed 1, 0, 1 and 1 after pseudo-code modulation (DSSS). That's why this technique of modulation is called as Code Division Multiplexing, and access to one code division multiplexed line, against permission, is called as CDMA.

7.7.3.2 DMUX

At the de-multiplexer side, firstly, the multiplexed sequence (2, −2, 2, 2) goes to 4 different dispreading/demodulator block. The four demodulators are synchronized with proper carrier sequences in order, i.e., seq0 (1111), seq1 (1100), seq2 (1010) and seq3 (1001) respectively.

Therefore, the results of 4 demodulators (analog bit by bit product) would be

$$\text{Ch0\_DMOD} \rightarrow (+1, +1, +1, +1)^{*}\ (2, -2, 2, 2) = (+2, -2, +2, +2)$$
$$\text{Ch0\_DMOD} \rightarrow (+1, +1, -1, -1)^{*}\ (2, -2, 2, 2) = (+2, -2, -2, -2)$$
$$\text{Ch0\_DMOD} \rightarrow (+1, -1, +1, -1)^{*}\ (2, -2, 2, 2) = (+2, +2, +2, -2)$$
$$\text{Ch0\_DMOD} \rightarrow (+1, -1, -1, +1)^{*}\ (2, -2, 2, 2) = (+2, +2, -2, +2)$$

Next, all the analog values are summed up and normalized to get the transmitted exact bit.

This procedure has a nice advantage that, if even a single digit is erroneous in the carrier chip sequence, the de-multiplexing cannot be done properly, because, spread spectrum; hence the CDMA is highly secure transmission.

References

1. Das, A., "Line Communication System – Telecommunication Switch Approach (2nd Ed.)", New Age International, 2010

2. Proakis, P., Salehi, M., "Fundamentals of Communication Systems", Pearson Education Inc, 2005
3. Farauzen, B.A., "Data Communication and Networking (4th Ed.)", Tata McGraw-Hill, 2006
4. Taub, H., Schilling, D.L. "Principles of Communication Systems", McGraw-Hill, 1986 (Original from the University of Michigan)
5. Chakraborty, N.B., Datta, A.K., "An Introduction to the Principles of Digital Communications", New Age International Pvt Ltd, 2007
6. Haykin, S., "Communication Systems (4th Ed.)", John Wiley and Sons, 2001
7. Kasami, T., "Weight Distribution Formula for Some Class of Cyclic Codes", Tech. Rep. No. R-285, April 1966, University of Illinois
8. Gold, R., "Optimal Binary Sequences of Spread Spectrum Multiplexing", *IEEE Transactions on Information Theory*, Vol. IT-13, October 1967, pp. 619–621

Chapter 8
Information Theory

8.1 Introduction

Psychologists look at the notion of 'information' in a nice and unique way. Let's consider a game of question answer with only 'Y' for 'yes' (positive) and 'N' for 'no' (negative) reply. A sequence of alphabets is given as shown in Fig. 8.1a. The letters are arranged chronologically and each is having equal probability, i.e. 1/8. Player 2 is to memorize any letter from the array. Initially no information about the choice is there with layer 1. The player 1 picks up one letter and asks 'Is it the chosen letter of player 2?' player 2 responds in Y or N. If Player 1 picks up letters in linear arrangement one by one, (linear search), the maximum number of information he has to know, i.e. number of times the same question is to be asked is the length of the array. Otherwise, player 1 may mark a demarcation line in between the array and ask intelligently, 'Is the chosen letter placed on the left side?' by this way, using a binary search, the maximum number of questions is to be answered is 3 for an array of 8 elements. Therefore, the amount of information is 3.

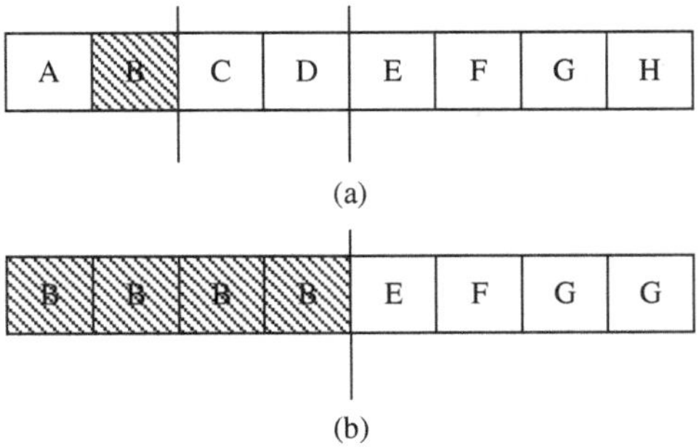

Fig. 8.1 Psychometric concept of information

As the probability of character B (the shaded B) is 1/8,

$$P = \frac{1}{8} = 2^{-3} \tag{8.1}$$

A. Das, *Digital Communication*, Signals and Communication Technology,
DOI 10.1007/978-3-642-12743-4_8,

It is clear that, the amount of information is

$$I = -\log_2\left(2^{-3}\right) = 3 \tag{8.2}$$

Next, let's consider another case as shown in Fig. 8.1b. There, the probability of B is 4/8, i.e., ½. And t is obvious that, the minimum number of questions to be asked is 1. Here also, probability is given by

$$P = \frac{1}{2} = 2^{-1} \tag{8.3}$$

and it is clear that, the amount of information is

$$I = -\log_2\left(2^{-1}\right) = 1 \tag{8.4}$$

Combining Eqs. (8.2) and (8.4), we get the measure of information as

$$I = -\log_2(P) = \log_2\frac{1}{P} \tag{8.5}$$

The concept of information and measure of information in accordance with the views of psychologists perfectly matches with the conjecture of Shanon in his revolutionary paper on 'A theory of communication' published in the *Bell System Technical Journal*, 1948 [1]. There also he defined the information logarithmically and average information as

$$\sum_{\forall i} p_i I_i, \text{ i.e.,} -\sum_{\forall i} p_i \log p_i.$$

Information theory deals with three basic concepts:

(i) The measure of source information: Evaluation of the rate at which source generates information.
(ii) The information capacity of channel: It determines the maximum rate at which reliable information transmission is possible over a given channel with an arbitrary small error.
(iii) Coding: A scheme for efficient utilization of the information capacity of the channel.

The combination of the above three concepts forms plenty of theorems and forms the entire discipline namely, *information theory*. In summary,

> If the rate of information from a source does not exceed the capacity of a given communication channel, then there exists a coding technique such that the information can be transmitted over the channel with arbitrary small errors, despite the presence of noise.

This implies that with the help of suitable coding scheme, it is possible to transmit signals from the source over a noisy channel almost error free.

8.2 Entropy

A communication system not always deals with a single message, moreover with all possible messages. Hence, although the instantaneous information flows corresponding to individual messages from a source may be erratic, we may describe the source in terms of *average information per message*, known as *entropy* of the source [3, 4].

Let's consider a situation where a source is generating M different messages, $m_1, m_2, \ldots, m_M$ with respective probabilities of occurrences $P_1, P_2, \ldots, P_M$. It's also assumed that, in a long time interval L, messages have been generated so that L>>M. Therefore, the number of m_1 message = P_1L.

The amount of information in message m_1 =log $\frac{1}{P_1}$. Thus the total amount of information in all m_1 messages= $P_1L\log\frac{1}{P_1}$.

The total amount of information in all L messages will then be

$$I_t = P_1L\log\frac{1}{P_1} + P_2L\log\frac{1}{P_2} + \ldots\ldots + P_ML\log\frac{1}{P_M}.$$

The average information per message or entropy will then be

$$\begin{aligned} H = \frac{I_t}{L} &= P_1\log\frac{1}{P_1} + P_2\log\frac{1}{P_2} + \ldots\ldots + P_M\log\frac{1}{P_M} \\ &= \sum_{i=1}^{M} P_i\log\frac{1}{P_i} = -\sum_{i=1}^{M} P_i\log P_i \end{aligned} \tag{8.6}$$

If there is a only a single message, P_1=1, then

$$H = P_1\log\frac{1}{P_1} = 1\log 1 = 0$$

Thus, it is inferred that, in case of single possible message, the reception conveys no information.

For a binary system, the possible messages are m_1 and m_2 with respective probabilities P_1 and P_2 with relation $P_1 + P_2 = 1$; Let, P_1=P, therefore $P_2 = (1-P)$. Therefore, the entropy is

$$\begin{aligned} H &= P_1\log\frac{1}{P_1} + P_2\log\frac{1}{P_2} \\ &= P\log\frac{1}{P} + (1-P)\log\frac{1}{(1-P)} \end{aligned} \tag{8.7}$$

The condition for maximum entropy can be obtained by differentiating H with respect to P,

$$\begin{aligned}\frac{dH}{dP} &= 0 = -\ln 2 - \log P + \ln 2 + \log(1-P)\\ &\Rightarrow \log P = \log(1-P)\\ &\Rightarrow P = 1-P\\ &\Rightarrow P = 1/2\end{aligned} \tag{8.8}$$

Also we have

$$\frac{d^2H}{dP^2} = -\frac{1}{P} - \frac{1}{1-P} < 0 \tag{8.9}$$

Hence, H has a maxima at P=1/2.

$$\therefore\ H_{\max} = \frac{1}{2}\log 2 + \frac{1}{2}\log 2 = 1\ \text{bit/message} \tag{8.10}$$

From the Fig. 8.2, the H pattern with varying P is plotted.

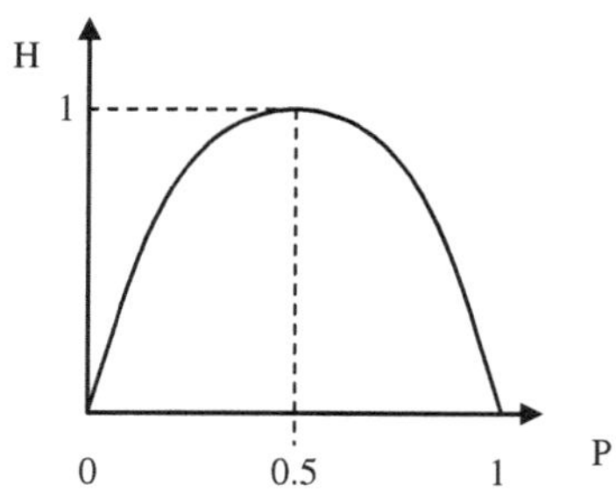

Fig. 8.2 Entropy of a binary source

So, for binary system (M=2), maximum entropy obtained when P=1/2, i.e., both the messages are equally likely. It can also be shown that, for M-ary transmission, the entropy is maximum when all the M messages are equally likely, i.e., P=1/M. Then the maximum entropy is

$$H_{\max} = \sum_{i=1}^{M} P_i \log\frac{1}{P_i} = M \times \frac{1}{M}\log M = \log M\,\text{bits/message} \tag{8.11}$$

The properties of entropy therefore are as follows:

(a) $\log M \geq H(x) \geq 0$
(b) $H(x) = 0$ if probabilities of (m−1) messages are 0 and probability of the rest one is 1.
(c) $H(x) = \log M$ when all probabilities are equal, i.e., $P_i = 1/M$ for all i.

From the three properties and the definition of entropy, it can be observed that, for certain cases, the entropy is zero. As the amount of uncertainty increases, the entropy value is also increased. With maximum entropy, the entropy is also maximized. Therefore, it is inferred that, *entropy is the measure of uncertainty*.

8.3 Rate of Information

In the previous section, entropy, H is defined as the average number of bits of information per message. Now, if a message source generates r messages per second, the rate of information is

$$R = rH \text{ bits/s} \tag{8.12}$$

Let's consider two sources with equal entropy H, generating r_1 and r_2 messages per second, respectively with $r_1 < r_2$. Therefore, rate of information is greater in the second source. It means, in a specific duration, source two will transmit more information than source one, placing greater demand on the communication channel. Hence, source is not only described by its entropy, but also by its rate of information.

Example 8.1 A CTCV signal is bandlimited to 4 kHz. The signal is quantized in 8 levels of PCM with probabilities 0.25, 0.2, 0.2, 0.1, 0.1, 0.05, 0.05 and 0.05. Calculate the entropy and the rate of information.

The Nyquist rate of sampling $= 4 \times 2 = 8$ kHz. Each sample is then quantized to one of the 8 levels.

The entropy is

$$\begin{aligned}\text{H} = &-(0.25\log 0.25 + 0.2\log 0.2 + 0.2\log 0.2 + 0.1\log 0.1 + 0.1\log 0.1 + \\ &0.05\log 0.05 + 0.05\log 0.05 + 0.05\log 0.05) = 2.74 \text{ bits/message}\end{aligned}$$

As the sampling frequency is 8 kHz, the message rate is 8,000 messages/s. Hence the rate of information is

$$\text{R} = \text{rH} = 8{,}000 \times 2.74 = 21{,}920 \text{ bits/s}$$

8.4 Information Sources

(a) *Continuous source*: The sources that produce continuous signals for transmission should be classified as continuous source [5, 3].
(b) *Discrete source*: The sources that produce signals only at discrete times are called discrete sources. A continuous analog signal after sampling is an example of a discrete source
(c) *Zero memory or memory less source*: When any message produced by a source is statistically independent of other messages produced before or after, the source is called a zero memory or memory less source. The probability of

any message is independent of the probabilities of other messages. In fact all examples discussed so far in this chapter are based on memory less source.

(d) *Extension of zero memory source:* After a memory less source has produced the messages, often successive two or more messages are combined together before transmission. These combined messages form a new set of symbols or messages. Consider a source, which can produce M messages namely m_1, m_2, m_3, ... etc. It randomly produces a string of messages, for example, $m_3-m_5-m_1-m_1-m_4-m_6-\ldots$ etc. Say, two successive messages are combined to make a symbol. Then the new set of messages will be m_{35}, m_{11}, m_{46}, ... etc., where m_{35} means the combination of m_3 and m_5.

Entropy of an extension of memory less source: If entropy of the basic memory less source is H and n nos. of successive messages are combined to form the nth order extension, then entropy of the nth order extended source is given by,

$$H(s^n) = n.\, H(s) \tag{8.13}$$

Example 8.2 A discrete memory less source (DMS) can produce four messages viz. m_1, m_2, m_3 and m_4 with probabilities 1/4, ½, 1/8 and 1/8. Determine the entropy of its 2nd order extension and show that it is twice the entropy of the basic source.

The problem has been illustrated in a tabular form in Table 8.1.

Table 8.1 Entropy of 2nd order extension of DMS

Basic source		2nd order extension		
Message	Probability	New message	Combination of	Joint probability
m_1	1/4	m_{11}	m_1 & m_1	1/4 × 1/4 = 1/16
m_2	1/2	m_{12}	m_1 & m_2	1/4 × 1/2 = 1/8
m_3	1/8	m_{13}	m_1 & m_3	1/4 × 1/8 = 1/32
m_4	1/8	m_{14}	m_1 & m_4	1/4 × 1/8 = 1/32
		m_{21}	m_2 & m_1	1/2 × 1/4 = 1/8
		m_{22}	m_2 & m_2	1/2 × 1/2 = 1/4
		m_{23}	m_2 & m_3	1/2 × 1/8 = 1/16
		m_{24}	m_2 & m_4	1/2 × 1/8 = 1/16
		m_{31}	m_3 & m_1	1/8 × 1/4 = 1/32
		m_{32}	m_3 & m_2	1/8 × 1/2 = 1/16
		m_{33}	m_3 & m_3	1/8 × 1/8 = 1/64
		m_{34}	m_3 & m_4	1/8 × 1/8 = 1/64
		m_{41}	m_4 & m_1	1/8 × 1/4 = 1/32
		m_{42}	m_4 & m_2	1/8 × 1/2 = 1/16
		m_{43}	m_4 & m_3	1/8 × 1/8 = 1/64
		m_{44}	m_4 & m_4	1/8 × 1/8 = 1/64
H = 1.75 bits/message		H2 = 3.5 bits/message		

From Eq. (8.6), entropy of the basic source is,

$$\begin{aligned} H(s) &= \sum_{i=1}^{M} p_i \log \frac{1}{p_i} \\ &= \frac{1}{4}\log_2 4 + \frac{1}{2}\log_2 2 + \frac{1}{8}\log_2 8 + \frac{1}{8}\log_2 8 = 1.75\ \text{bits/message} \end{aligned} \tag{8.14}$$

When messages are produced at random and two successive messages are combined, $4^2 = 16$ combinations are possible. Joint probability of each combination is determined by multiplying the two individual probabilities. Second order entropy is calculated on these joint probabilities.

$$\begin{aligned} H_{n=2} &= \sum_{i=1}^{M} p_i \log \frac{1}{p_i} \\ &= \frac{1}{16}\log_2 16 + \frac{1}{8}\log_2 8 + \frac{1}{32}\log_2 32 + \frac{1}{32}\log_2 32 + \frac{1}{8}\log_2 8 \\ &\quad + \frac{1}{4}\log_2 4 + \frac{1}{16}\log_2 16 + \frac{1}{16}\log_2 16 + \frac{1}{32}\log_2 32 \\ &\quad + \frac{1}{16}\log_2 16 + \frac{1}{64}\log_2 64 + \frac{1}{64}\log_2 64 + \frac{1}{32}\log_2 32 \\ &\quad + \frac{1}{16}\log_2 16 + \frac{1}{64}\log_2 64 + \frac{1}{64}\log_2 64 = 3.5\ \text{bits/message} \end{aligned} \tag{8.15}$$

Thus it is established that by 2nd order extension, the entropy of the basic source is doubled. Hence, $H(s^2) = 2\,H(s)$ and similarly, $H(s^n) = n.H(s)$

(e) *Markov source:* When probability of occurrence of any message depends upon the occurrence of earlier one or more messages, the source is called a Markov source. If its probability depends upon earlier q messages, then it is called a qth order Markov source. In such cases, information content of any message is measured by its conditional probability after earlier messages.

Say, an event B occurs after the event A has occurred and probability of occurrence of B depends upon whether A has occurred immediately before, then the probability of occurrence of B after A is given by

$$p\,(B/A) = \frac{p(B,A)}{p(A)},$$

where p(B, A) is the joint probability of A p(A) & B and p(A) is the independent probability of event A.

In a 1st order Markov source if message m_2 occurs after message m_1 then information content of m_2 is given by,

$$I(m_2/m_1) = \log_2 \frac{1}{p(m_2/m_1)} = \log_2 \frac{p(m_1)}{p(m_1, m_2)} = \log_2 \frac{p_1}{p_1 p_2} \tag{8.16}$$

where p_1 and p_2 are independent probabilities of m_1 and m_2.

Plain English text is a good example of a Markov source. Probability of any letter depends upon the earlier letter/s. For example, probability of a 'u' after 'q' can be considered 100%. Hence, once a 'q' has occurred, a 'u' thereafter brings no new information.

Mathematically speaking, p(u/q) = 1. Hence, information content of 'u' after 'q' $= I_{(u/q)} = \log_2 \frac{1}{p(u/q)} = \log_2 \frac{1}{1} = 0$

8.5 Discrete Memoryless Channel (DMC)

8.5.1 Channel Representation

As we know, a communication channel may be defined as the medium through which the symbols flow from the transmitter to the receiver end. A *discrete memoryless channel* is a statistical model with an input random variable X and an output random variable Y, as shown in Fig. 8.3. At each unit duration, one symbol from input is accepted and one symbol at the output responds. In discrete channel, the number of alphabets in X and Y are both finite and in memoryless channel, the current output only depends on the current input not the delayed (latched) previous one.

According to the Fig. 8.2, the input X consist of x_i s with i=1, 2, ... m with a priory known probabilities $P(x_i)$ s respectively. Similarly, output Y consist of y_i s with i=1, 2, ... m. Each input-to-output path is indicated alongwith the conditional probability $P(y_j|x_i)$, where $P(y_j|x_i)$ is the conditional probability of obtaining output y_j given that the input is x_i, and is called a *channel transition probability.*

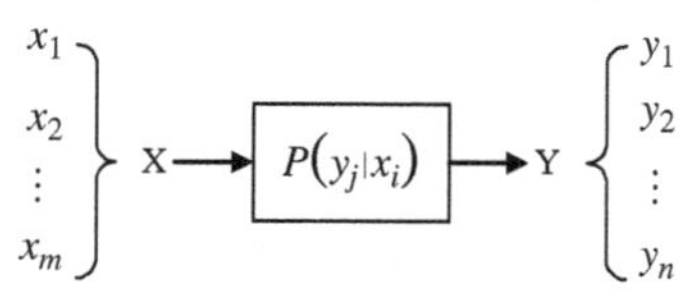

Fig. 8.3 Representation of discrete memoryless channel

8.5.2 The Channel Matrix

A channel is specified by a complete set of transition probabilities. According to the Fig. 8.2, the channel is specified by the matrix of transition probabilities$[P(Y|X)]$. The matrix is given by

$$[P(Y|X)] = \begin{bmatrix} P(y_1|x_1) & P(y_1|x_1)\ldots & P(y_1|x_1) \\ P(y_1|x_1) & P(y_1|x_1)\ldots & P(y_1|x_1) \\ \vdots & \vdots & \ldots & \vdots \\ P(y_1|x_1) & P(y_1|x_1)\ldots & P(y_1|x_1) \end{bmatrix} \quad (8.17)$$

The matrix $[P(Y|X)]$ is called the *channel matrix.*

Since each input to the channel results in some output, each row of the channel matrix must sum to unity. This means that

$$\sum_{j=1}^{n} P(y_j|x_i) = 1, \forall i \quad (8.18)$$

Now, if the input probability P(X) is represented by the row matrix, then we have

$$[P(X)] = [P(x_1) \qquad P(x_2)\ldots P(x_m)] \quad (8.19)$$

Also, the output probability P(Y) is represented by the row matrix as follows:

$$[P(Y)] = \left[P(y_1) \qquad P(y_2)\ldots P(y_n)\right] \quad (8.20)$$

then

$$[P(Y)] = [P(X)]\,[P(Y|X)] \quad (8.21)$$

Now, P(X) is represented as a diagonal matrix, then

$$[P(X,Y)] = [P(X)]\,[P(Y|X)] \quad (8.22)$$

The matrix P(X, Y) is the *joint probability matrix.*

8.6 Special Channels

There are some special channels also which are neither discrete nor exactly continuous. These channels are having their own channel matrices.

8.6.1 Lossless Channel

In lossless channels, no source information is lost due to transmission [2, 5]. This channel descriptor or channel matrix holds only one non-zero element each column, as shown in the Eq. (8.23) and Fig. 8.4.

$$[P(Y|X)] = \begin{bmatrix} \frac{3}{4} & \frac{1}{4} & 0 & 0 & 0 \\ 0 & 0 & \frac{1}{3} & \frac{2}{3} & 0 \\ 0 & 0 & 0 & 0 & 1 \end{bmatrix} \tag{8.23}$$

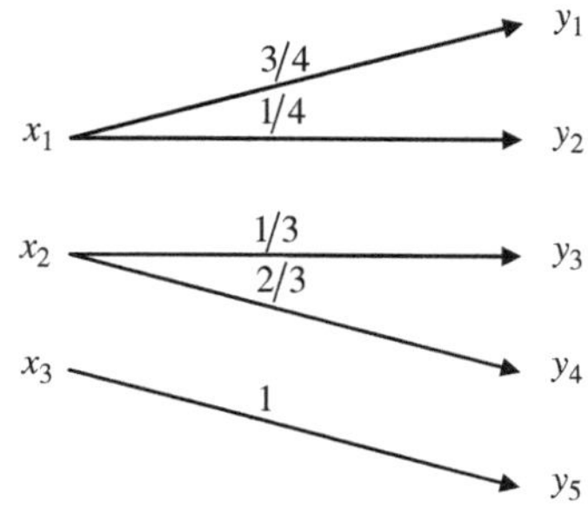

Fig. 8.4 Lossless channel

8.6.2 Deterministic Channel

Deterministic channel matrix holds only one and unique non-zero element at each row as shown in the Fig. 8.5. The channel matrix is given by

$$[P(Y|X)] = \begin{bmatrix} 1 & 0 & 0 \\ 1 & 0 & 0 \\ 0 & 1 & 0 \\ 0 & 1 & 0 \\ 0 & 0 & 1 \end{bmatrix} \tag{8.24}$$

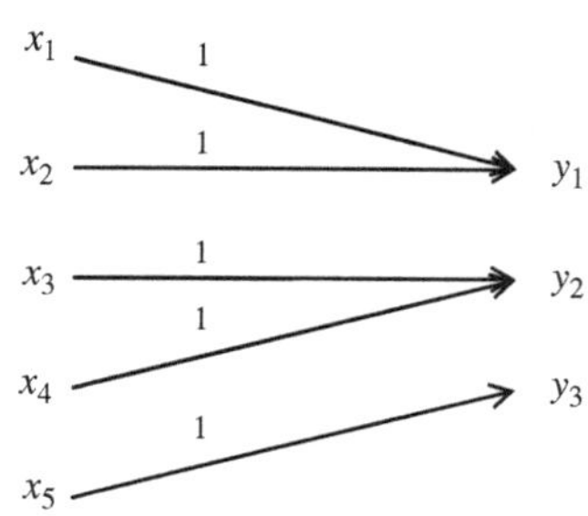

Fig. 8.5 Deterministic channel

Since, the Eq. (8.24) says, summation of elements of each row must be unity, and in deterministic channel, each row holds only one element, therefore the element itself must be unity. The physical interpretation of the matrix is, ‘if a symbol is transmitted it must be received’.

8.6.3 Noise-Less Channel

The channel which is both lossless and deterministic is called as noise-less channel (Fig. 8.6). Obviously, the channel matrix here is an identity matrix as shown in the following equation.

$$[P(Y|X)] = \begin{bmatrix} 1 & 0 & 0 & 0 \\ 0 & 1 & 0 & 0 \\ 0 & 0 & 1 & 0 \\ 0 & 0 & 0 & 1 \end{bmatrix} \tag{8.25}$$

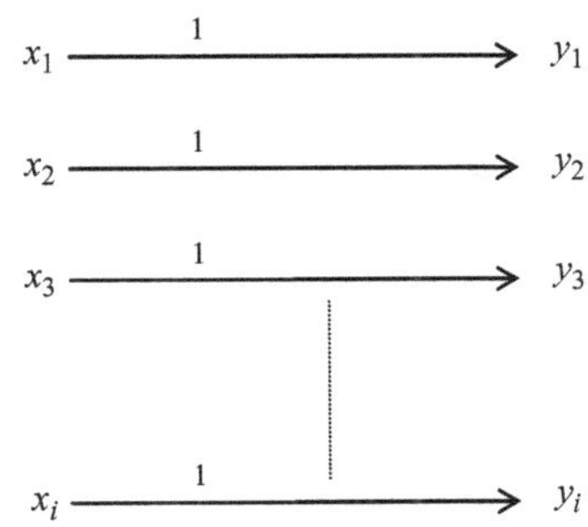

Fig. 8.6 Noise-less channel

8.6.4 Binary Symmetric Channel (BSC)

Binary channels have two inputs ($x_1 = 0, x_2 = 1$) and two outputs ($y_1 = 0, y_2 = 1$). The channel becomes symmetric when the probability, p of receiving 0 when 1 is transmitted is same as probability of receiving 1 when 0 is transmitted. The channel matrix is then defined by

$$[P(Y|X)] = \begin{bmatrix} 1-p & p \\ p & 1-p \end{bmatrix} \tag{8.26}$$

The channel is therefore represented as shown in Fig. 8.7.

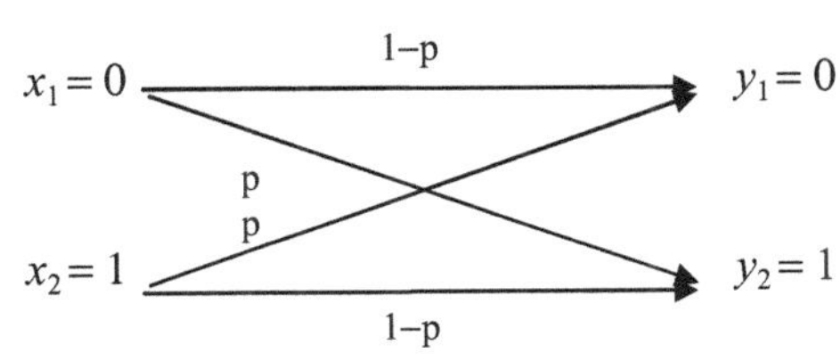

Fig. 8.7 Binary symmetric channel

8.6.4.1 Saturated or Stable Probability of Error for Cascaded BSC Channel

With a nice example and program, it can be shown that, if N numbers of BSC channels are cascaded, after passing through a point of BSC unit, all the error probabilities will results in a saturated value of ½. In the Fig. 8.8 it is shown that, with two different initial values of probability of errors (p_{01} or p_{10}), parametric graphs are obtained with cascaded 10 units of BSC. The MATLAB code given in Sect. 8.10 can help more to understand the convergence of the error probability to ½. It is shown that, after traversing through 6 iterations, i.e., 6 units of BSC, the initial values of p_{01}=0.3 and 0.65, both the case saturates to 0.5.

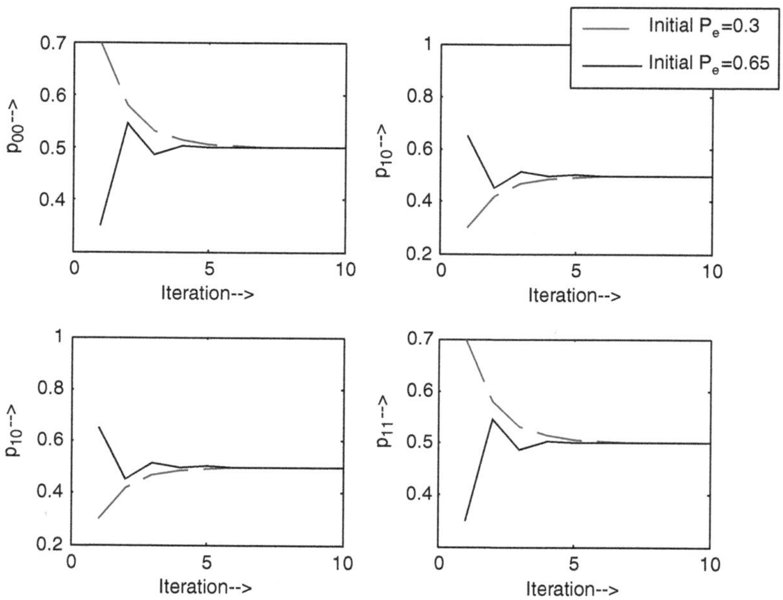

Fig. 8.8 Probability or error is converging in cascaded BSC

8.6.4.2 Probability Model of Erroneous Detection in Cascaded BSC

Cascaded combination of BSC is considered, where error can occur in two ways, one by right followed by wrong detection and other wrong followed by right detection (Fig. 8.9). Right followed by right and wrong followed by wrong detection results in right detection of message symbol.

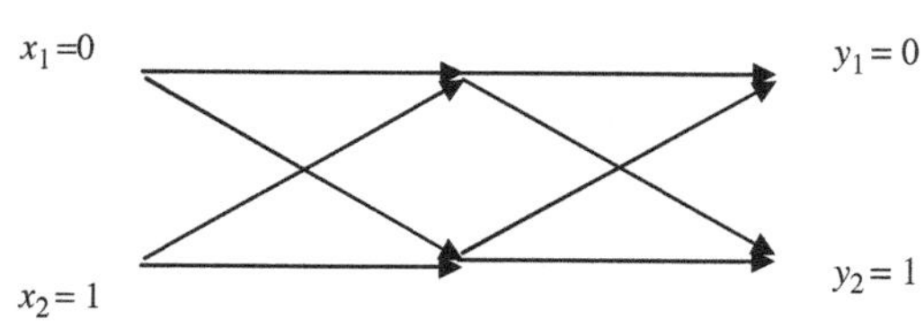

Fig. 8.9 Cascaded BSC

Repeater1	Repeater2	Rx
C	C	C
C	W	W
W	C	W
W	W	C

Therefore, the probability of erroneous detection is ½.

The number ½ is really interesting. The number can be expressed as

$$\frac{1}{2} = 2 \times \frac{1}{2} \times \frac{1}{2} = {}^{2}C_{1} \times \left[\frac{1}{2}\right]^{1} \times \left[\frac{1}{2}\right]^{(2-1)} \tag{8.27}$$

For a 3 repeater cascade, the table would be as follows:

Repeater1	Repeater2	Repeater3	Rx
W	W	W	W
W	W	C	C
W	C	W	C
W	C	C	W
C	W	W	C
C	W	C	W
C	C	W	W
C	C	C	C

Here, the probability of erroneous detection is 4/8=1/2. This is a priory probability. But, out of these 4 wrong (W) reception, in 3 cases there are 1 W repeater, therefore, probability is

$$\frac{3}{4} \times \frac{1}{2} = 6 \times \frac{1}{2} \times \frac{1}{4} = {}^{3}C_{1} \times \left[\frac{1}{2}\right]^{1} \times \left[\frac{1}{2}\right]^{(3-1)} \tag{8.28}$$

Hence, Eqs. (8.27) and (8.28) suggests the distribution of error probability model will be

$$P_e = {}^{n}C_{x} p^{x} (1-p)^{n-x} \tag{8.29}$$

This is clearly a *binomial distribution*.

8.7 Mutual Information

Before receiving the transmitted symbol x_j at the receiver, the state of knowledge at the receiver is the probability that x_j would be selected for transmission. This is known as *a-priory probability* $p\left(x_j\right)$. After receiving the symbol y_k the state of knowledge concerning x_j is the conditional probability $p\left(x_j\,|y_k\right)$, known as *a posteriori probability.* Thus before and after reception, the uncertainty is $-\log p\left(x_j\right)$ and $-\log p\left(x_j\,|y_k\right)$, respectively.

The information gained about x_j by the reception of y_k is the net reduction in the uncertainty, and is known as *mutual information* $I\left(x_j; y_k\right)$.

Thus, $I\left(x_j; y_k\right) =$ initial uncertainty-final uncertainty

$$
\begin{aligned}
&= -\log p\left(x_j\right) - \left\{-\log p\left(x_j\,|y_k\right)\right\} \\
&= \log \frac{p\left(x_j\,|y_k\right)}{p\left(x_j\right)} \\
&= \log \frac{p\left(x_j, y_k\right)}{p\left(x_j\right) p\left(y_k\right)} \\
&= \log \frac{p\left(y_k|x_j\right)}{p\left(y_k\right)} = I\left(y_k; x_j\right)
\end{aligned}
\tag{8.30}
$$

Equation (8.30) says, '*mutual information is symmetric*'.

Next, the average of mutual information, i.e., the entropy corresponding to mutual information is given by

$$
\begin{aligned}
I(X;Y) &= \bar{I}\left(x_j; y_k\right) \\
&= \sum_{j=1}^{m}\sum_{k=1}^{n} p\left(x_j, y_k\right) I\left(x_j; y_k\right) \\
&= \sum_{j=1}^{m}\sum_{k=1}^{n} p\left(x_j, y_k\right) \log \frac{p\left(x_j\,|y_k\right)}{p\left(x_j\right)} \\
&= \sum_{j=1}^{m}\sum_{k=1}^{n} p\left(x_j, y_k\right) \left[\log p\left(x_j\,|y_k\right) - \log p\left(x_j\right)\right] \\
&= -\sum_{j=1}^{m}\sum_{k=1}^{n} p\left(x_j, y_k\right) \log p\left(x_j\right) - \left[-\sum_{j=1}^{m}\sum_{k=1}^{n} p\left(x_j, y_k\right) \log p\left(x_j\,|y_k\right)\right] \\
&= -\sum_{j=1}^{m}\left[\sum_{k=1}^{n} p\left(x_j, y_k\right)\right] \log p\left(x_j\right) - H(X/Y) \\
&= -\sum_{j=1}^{m} p\left(x_j\right) \log p\left(x_j\right) - H(X/Y)
\end{aligned}
$$

$$= H(X) - H(X/Y) \tag{8.31}$$

$$= H(Y) - H(Y/X) \tag{8.32}$$

The above two equations signifies that, the mutual information does not depend at all upon the input or output symbol samples, it is a property of whole communication system, specially a communication channel.

8.8 Channel Capacity

The channel capacity is defined as the maximum amount of mutual information. Actually, the transmission efficiency is the comparison between actual rate of information transmission and upper bound of information transmission rate through the channel. Therefore, the channel capacity is defined as

$$C = \max I(X;Y) = \max\left[H(X) - H(X/Y)\right] \tag{8.33}$$

The channel efficiency is defined as

$$\eta = \frac{I(X;Y)}{\max I(X;Y)} = \frac{I(X;Y)}{C} \tag{8.34}$$

8.8.1 Gaussian Channel: Shanon-Hartley Theorem

The study of Gaussian channel[1] is really important because, the Mutual information obtained from Gaussian channel provide a lower bound on the performance of the system with the non-Gaussian channel. Thus, if a particular channel encoder is used for a Gaussian channel, giving probability of error P_e, then for a non-Gaussian channel another channel encoder can be designed so that, the probability of error would be less than P_e.

For a Gaussian channel,

$$p(x) = \frac{1}{\sqrt{2\pi\sigma^2}} e^{-\frac{x^2}{2\sigma^2}} \tag{8.35}$$

[1]Channels with Gaussian noise characteristics are known as Gaussian channel.

Hence, channel entropy is

$$
\begin{aligned}
H(x) &= -\int_{-\infty}^{\infty} p(x) \log p(x)\, dx \\
&= \int_{-\infty}^{\infty} p(x) \log \sqrt{2\pi\sigma^2} dx + \int_{-\infty}^{\infty} p(x) \log e^{\frac{x^2}{2\sigma^2}} dx \\
&= \log \sqrt{2\pi e\sigma^2} \text{ bits/message}
\end{aligned} \tag{8.36}
$$

Now, the signal is band-limited to ω Hz, then it may be uniquely specified taking at least 2ω samples.[2] Hence the rate of information transmission is

$$
\begin{aligned}
&R(x) = 2\omega H(x) = 2\omega \log \sqrt{2\pi e\sigma^2} = \omega \log \left(\sqrt{2\pi e\sigma^2}\right)^2 \\
&\therefore\ R(x) = \omega \log \left(2\pi e\sigma^2\right)
\end{aligned} \tag{8.37}
$$

If p(x) is band-limited Gaussian noise with an average noise power N, then

$$R(n) = R(x) = \omega \log (2\pi eN) \quad \left[\because\ \sigma^2 = N\right] \tag{8.38}$$

Now, let us consider the case of continuous data transmission through a noisy channel. If the received signal is composed of a transmitted signal x and a noise n, then the joint entropy of the source and noise is given by

$$R(x, n) = R(x) + R(n/x) \tag{8.39}$$

In practice, transmitted symbol and noise are independent. Therefore,

$$R(x, n) = R(x) + R(n) \tag{8.40}$$

Considering the received signal y as sum of transmitted signal x and noise n,

$$
\begin{aligned}
H(x, y) &= H(x, n) \\
H(y) + H(x/y) &= H(x) + H(n) \\
R(y) + R(x/y) &= R(x) + R(n)
\end{aligned} \tag{8.41}
$$

The rate at which the information is received from a noisy channel is

$$
\begin{aligned}
R &= R(y) - R(x/y) \\
R &= R(y) - R(n)
\end{aligned} \tag{8.42}
$$

[2] Nyquist rate of sampling.

The channel capacity is

$$C = Max\,[R] = Max\left[R(y) - R(n)\right] \text{ bits/s} \tag{8.43}$$

Since, R(n) is independent of x(t), maximizing R requires maximizing R(y).

Let, the power of signal and noise are S and N respectively within a bandwidth ω of the channel [2]. The received signal will now have an average power (S+N). R(y) is maximum when y(t) is also a Gaussian random process because noise is assumed to be Gaussian. Thus, the entropy from Eq. (8.38) on a per second basis is

$$R(y) = \omega \log\{2\pi e\,(S+N)\} \text{ bits/s} \tag{8.44}$$

And the entropy of the noise is given by

$$R(n) = \omega \log\{2\pi e N\} \text{ bits/s} \tag{8.45}$$

Then, the channel capacity can be expressed directly as

$$\begin{aligned} C &= \max\left[R\,(y) - R\,(n)\right] \\ &= \omega\left[\log 2\pi e\,(S+N) - \log 2\pi e N\right] \\ &= \omega\left[\log \frac{(S+N)}{N}\right] \\ &= \omega \log\left[1 + \frac{S}{N}\right] bits/s \end{aligned} \tag{8.46}$$

The above equation is the *Shanon-Hartley theorem.*

Now, to be specific, if the two sided power spectral density (PSD) of noise is $\eta/2$,

$$C = \omega \log\left[1 + \frac{S}{\eta\omega}\right] \text{bits/s} \tag{8.47}$$

8.9 Entropy Coding

A source is assumed which is generating a number of messages. If the number is N, it is easily understandable that, the length of the code for constant length coding will be $\lceil \log_2 N \rceil$. In practice, assigning same length to each message is a misuse. As for an example, if the length comes 10, for a frequently occurred message, we are assigning 10 bits at the same time, for a rarely occurred message we are assigning 10 bits. If there were a system which can assign the length of the code word depending upon the probability of occurrence in a particular context (Locally) or in generalized context (Globally), then the average number of code could be reduced, i.e., the word length can be used more efficiently.

The design of such variable length coding such that its average code length approaches the *entropy* of DMS is often referred as *entropy coding*. There the average length, $\overline{L}$ of the code is given by

$$\overline{L} = \sum_{\forall i} p_i L_i \tag{8.48}$$

where, L_i and p_i are the length and probability of ith code. Hence, the amount of *efficiency the code* can obtain is the ratio of entropy to the average code length, given by

$$\eta = H(X) / \overline{L} \tag{8.49}$$

The *code redundancy* is defined as

$$\gamma = 1 - \eta \tag{8.50}$$

Two examples of entropy coding are presented in this section.

8.9.1 Shanon-Fano Coding

The following steps of algorithm, known as Shanon-Fano algorithm can generate an efficient code vector.

1. List of the source symbols (x_is) are arranged in order of decreasing probability (p_is).
2. Partition the set into two sets that are close to equiprobable sets.
3. Continue this process of partitioning until further partitioning is not possible.

An example of Shanon-Fano coding is shown in the Table 8.2. It's to be noted that, a small ambiguity may arise choosing equiprobable messages and hence code length. But it does not affect on the average length of code, i.e., the code efficiency.

Now, the mean length of the code words is

$$\begin{aligned}\overline{L} = \sum p_i L_i &= 0.48 \times 1 + 0.15 \times 3 + 0.10 \times 3 + 0.10 \times 4 \\ &+ 0.07 \times 4 + 0.05 \times 4 + 0.03 \times 5 + 0.03 \times 5 = 2.36\end{aligned} \tag{8.51}$$

Table 8.2 Example Shanon-Fano coding

x_is	p_is	Order of selection: 1st	2nd	3rd	4th	5th	Shanon-Fano code	Constant length code
A	0.48	0					0	0 0 0
B	0.15	1	0	0			1 0 0	0 0 1
C	0.10			1			1 0 1	0 1 0
D	0.10		1	0	0		11 0 0	0 1 1
E	0.07				1		1 1 0 1	1 0 0
F	0.05			1	0		1 1 1 0	1 0 1
G	0.03				1	0	1 1 1 1 0	1 1 0
H	0.02					1	1 1 1 1 1	1 1 1

Entropy is

$$\begin{aligned} H(X) = -\sum p_i \log p_i = &-(0.48 \log_2 0.48 + 0.15 \log_2 0.15 \\ &+ 2 \times 0.1 \log_2 0.1 + 0.07 \log_2 0.07 + 0.05 \log_2 0.05 \\ &+ 0.03 \log_2 0.03 - 0.02 \log_2 0.02) = 2.33 \end{aligned} \tag{8.52}$$

Efficiency of the code is defined as

$$\eta = H(X)/\bar{L} = 0.987 \Rightarrow 98.7\% \tag{8.53}$$

Redundancy of the code is

$$\gamma = 1 - \eta = 1.3\% \tag{8.54}$$

8.9.2 Huffman Coding

Optimum code generation is named as Huffman coding. This code has the maximum efficiency and hence minimum redundancy. The encoding procedure is as the following step suggests.

1. List of the source symbols (x_is) are arranged in order of decreasing probability (p_is).
2. Combine the probabilities of the two symbols having the lowest probabilities, and reorder the resultant probabilities, this step is called *reduction 1*. The same procedure is repeated until there remain two ordered probabilities.

Table 8.3 Example of Huffman coding

x_i s	p_i s	p_i s of reduced order set								
		Code	1st		2nd		3rd		4th	
A	0.48	0	0.48		0.48		0.48	0	0.48	0
B	0.15	11	0.15		0.15		0.22	10	0.37	1
C	0.10	101	0.10	101	0.12	100	0.15	11		
D	0.07	1000	0.07	1000	0.10	101				
E	0.03	10010	0.05	1001						
F	0.02	10011								

3. Start encoding with the last reduction, which consist of exactly two ordered probabilities. Assign 0 as the 1st digit in the codeword for all the source symbols associated with the 1st probability; assign 1 to the second probability.
4. Go backward and subsequently assign 0 and 1 to the second digit for the two probabilities that were combined in the previous reduction step, retaining all assignments made in Step 3.
5. Keep the regressing in this way until the 1st column is reached.

An example of Huffman coding is shown in Table 8.3.
Now, the mean length of the code words is

$$\begin{aligned}\overline{L} = \sum p_i L_i &= 0.48 \times 1 + 0.15 \times 2 + 0.10 \times 3 \\ &\quad + 0.07 \times 4 + 0.03 \times 5 + 0.02 \times 5 = 1.61\end{aligned} \tag{8.55}$$

8.10 MATLAB Code

8.10.1 Convergence of P_e in Cascaded BSC

```
% In CD: ch8_1.m
% Cascaded BSC
% Output: Fig. 8.8
% Programed by Apurba Das (Dec,'09)
N=input('Enter the number of units of BSC: ');
p=input('Enter the alue of error probability P_01: ')
a=[(1-p),p;p,(1-p)];
b=a;
c00(1)=(1-p);c01(1)=(p);c10(1)=(p);c11(1)=(1-p);
```

```
for i=2:N
    b=b*a;
    c00(i)=b(1,1);
    c01(i)=b(1,2);
    c10(i)=b(2,1);
    c11(i)=b(2,2);
    b=b;
end;
i=1:N;
subplot(221)
plot(i,c00,'k')
xlabel('Iteration->');
ylabel('p_{00}->');
grid;
subplot(222)
plot(i,c01,'k')
xlabel('Iteration->');
ylabel('p_{01}->');
grid;
subplot(223)
plot(i,c10,'k')
xlabel('Iteration->');
ylabel('p_{10}->');
grid;
subplot(224)
plot(i,c11,'k')
xlabel('Iteration->');
ylabel('p_{11}->');
grid;
```

References

1. Shanon, C.E., "A Theory of Communication", *Bell System Technical Journal*, Vol. 27, July, October 1948, pp. 379–423, 623–656
2. Bedrosian, E., "Spectral Conservation by Efficient Channel Utilization", *Rand Corporation Report*, WN-9255-ARPA, Contract DAHC-15-73-C-0181. Santa Monica, California, October, 1975
3. Sharma, S., "Communication Systems – Analog and Digital." S. K. Kataria and Sons, 2009
4. Korn, I., "Principles of Coherent Communications", McGraw-Hill Book Co, 1966
5. Chakraborty, N.B., Datta, A.K., "An Introduction to the Principles of Digital Communications", New Age International Pvt Ltd, 2007

Chapter 9
Error Control Coding

9.1 Introduction

In communication a message must have to travel a distance before reaching the destination. In reality no message can travel this distance without undesirable error and hence comes the concept of error control. Coding technique can be fruitfully applied in error controlling. In coding, a message in encoded into one of the predefined code words. The working of error control code can be better understood by the following example.

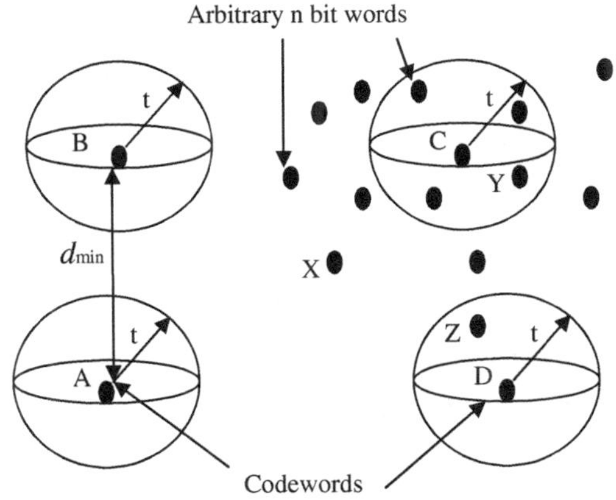

Fig. 9.1 Code classification and region of definition

In a particular decoder, a lookup table contains the set of valid code words. In Fig. 9.1, each valid codeword is represented by a dot surrounded by a sphere. The radius t of the sphere indicates the number of errors the code can correct. Now let us consider, codeword C has been transmitted. In the receiver, if C is received then no error has been occurred. If Y is received, then the receiver detects it as an error and as Y is in C's sphere, it will correct the error to give C, hence the decoding is successful. This is the case of detectable and correctable error pattern. If Z is received, then it will be corrected as D and as the final output is not C, the decoding fails. This is called false code acceptance. If D is received, then no error will be detected by

A. Das, *Digital Communication*, Signals and Communication Technology,
DOI 10.1007/978-3-642-12743-4_9, © Springer-Verlag Berlin Heidelberg 2010

the decoder and the decoding fails unfortunately. This error is undetectable to the decoder. This is also a case of false code acceptance. If X is received, the decoder detects the error, but can't correct it as it is not inside any sphere. So decoding fails. This is the case of detectable but not correctable error pattern. Based on this explanation the error control code can be divided in 3 categories [1, 4, 5].

1. Undetectable
2. Detectable but not correctable
3. Detectable and correctable

9.2 Scope of Coding

Now a days the role of coding has expanded tremendously in various fields and today coding can do the following:

Reduce the occurrence of undetected errors: This is the most important advantage of coding. Today's error control codes are so effective that the occurrence of undetectable errors is eliminated for all practical purposes.

Reduce the cost of communication systems: Coding can significantly reduce the power needs of transmitter especially of satellite reducing the communication cost because signals received at thermal noise level can still be recovered correctly.

Overcome jamming: Coding is one of the most effective techniques for reducing the effects of jamming.

Eliminate interference: Error control coding is very effective to reduce the effect of unintentional interference.

Despite all these features, there are some limitations of coding too. One of these is the modulation technique of communication system. Coding must simultaneously with the choice of modulation technique for the channel. Even the most powerful codes cannot overcome the consequences of a poor modulation choice.

Another limitation is the inability to achieve Shannon's lower bound.

• Forward Error Correction

If we compare the sphere of code wards A and B in figure, we will see that the error correcting capability of a code is given by $d_{\min} = 2t + 1$ (this is the minimum separation that prevents overlapping spheres). In other words, a code with $d_{\min} = 3$ or 5 can correct all patterns of 1 or 2 errors.

9.3 Linear Block Code

In FEC (Forward Error Correction) method the channel encoder automatically adds digits to the message that can be checked in the receiver side to find out errors and correct it. The message with the additional digits is now called code. There are two categories of this code:

1. Block Code
2. Convolutional Code

The continuous message from the source is divided into segments of n number of bits. These segments are called blocks. The channel encoder then adds another k bits to the blocks to form the codes. These k check bits are derived from the message bits and are added to them. The $(n + k)$ bits of the encoder output are called code word. In the coding techniques where the message bits appear at the beginning of the code word are called *Systematic Codes* (Fig. 9.2)

Fig. 9.2 Block code format

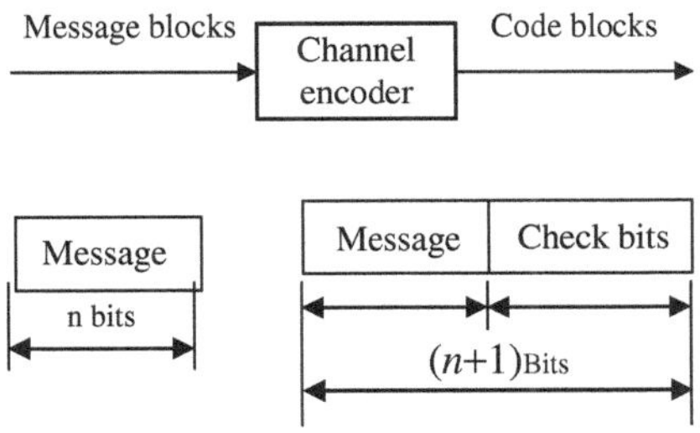

If sum of any two code words gives another code word of a block code then the block code is said to be linear, i.e. Linear Block Code.

$$c_i + c_j = c_k$$

where c_i, c_j, c_k are code words of a block code.

9.3.1 Coding Technique Using Generator Matrix

In systematic linear block code, the code word is the information bits followed by the check bits. These check bits are generated from the message bits according to some predefined rule. Let us consider a systematic linear block code with k no. of message bits and n−k no. of check bits. So the code word is c_j where j=1, 2, 3…., n. Now for j=1 to k, c_j=d_j where d_j is the message bit. The check bits are calculated using the following rule:

$$
\begin{aligned}
C_{k+1} &= p_{11}d_1 \oplus p_{21}d_2 \oplus \cdots \oplus p_{k1}d_k \\
C_{k+2} &= p_{12}d_1 \oplus p_{22}d_2 \oplus \cdots \oplus p_{k2}d_k \\
&\cdots\cdots\cdots\cdots\cdots\cdots\cdots\cdots \\
C_n &= p_{1,n-k}d_1 \oplus p_{2,n-k}d_2 \oplus \cdots \oplus p_{k,n-k}d_k
\end{aligned}
\tag{9.1}
$$

where $p_{i,j}$ is 0 or 1 such that C_k is 0 or 1. The above rule can be written in matrix from giving a matrix equation:

$$
[C_1 C_2 \cdots C_n] = [d_1 d_2 \cdots d_k]
\begin{bmatrix}
1 & 0 & 0 & 0 & \cdots & 0 & \vdots & p_{11} & p_{12} & \cdots & p_{1,n-k} \\
0 & 1 & 0 & 0 & \cdots & 0 & \vdots & p_{21} & p_{22} & \cdots & p_{2,n-k} \\
0 & 0 & 1 & 0 & \cdots & 0 & \vdots & p_{31} & p_{32} & \cdots & p_{3,n-k} \\
\cdots & \cdots & \cdots & \cdots & \cdots & \cdots & \vdots & \cdots & \cdots & & \cdots \\
0 & 0 & 0 & 0 & \cdots & 1 & \vdots & p_{k1} & p_{k2} & \cdots & p_{k,n-k}
\end{bmatrix}
\tag{9.2}
$$

$$
\text{Or } C = DG \tag{9.3}
$$

Here G is the $k \times n$ matrix on the RHS of the above equation. This matrix G is called the generator matrix. The general form of G is $G = \left[I_k \vdots P \right]_{k \times n}$. A block code generated with $k \times n$ order generator matrix is called a (n, k) block code.

Example 9.1 Determine the set of code words for the (6,3) code with generator matrix

$$
G = \begin{bmatrix} 1\,0\,0\,0\,1\,1 \\ 0\,1\,0\,1\,0\,1 \\ 0\,0\,1\,1\,1\,0 \end{bmatrix}
$$

We need to consider the information words (0 0 1), (0 1 0), (0 1 1),. . .,(1 1 1). Substituting G and $i = \begin{pmatrix} 0\,0\,1 \end{pmatrix}$ into the equation gives the code word

$$
c = \begin{pmatrix} 0\,0\,1 \end{pmatrix} \begin{bmatrix} 1\,0\,0\,0\,1\,1 \\ 0\,1\,0\,1\,0\,1 \\ 0\,0\,1\,1\,1\,0 \end{bmatrix} = \begin{pmatrix} 0\,0\,1\,1\,1\,0 \end{pmatrix}
$$

And $i = \begin{pmatrix} 0\,1\,0 \end{pmatrix}$ gives the code word

$$
c = \begin{pmatrix} 0\,1\,0 \end{pmatrix} \begin{bmatrix} 1\,0\,0\,0\,1\,1 \\ 0\,1\,0\,1\,0\,1 \\ 0\,0\,1\,1\,1\,0 \end{bmatrix} = \begin{pmatrix} 0\,1\,0\,1\,0\,1 \end{pmatrix}
$$

Substituting the other information words, all the code words found and they are shown in Table 9.1.

Table 9.1 (6, 3) Linear block code example

i	c
(0 0 0)	(0 0 0 0 0 0)
(0 0 1)	(0 0 1 1 1 0)
(0 1 0)	(0 1 0 1 0 1)
(0 1 1)	(0 1 1 0 1 1)
(1 0 0)	(1 0 0 0 1 1)
(1 0 1)	(1 0 1 1 0 1)
(1 1 0)	(1 1 0 1 1 0)
(1 1 1)	(1 1 1 0 0 0)

9.3.2 Syndrome Decoding

Earlier we have seen that a generator matrix G can be expressed as $G = \left[I_k \vdots P \right]_{k \times n}$ where P is the k by n−k parity matrix. From the generator matrix, a parity check matrix can be constructed in following way [3]

$$H = \left[P^T \mid I_{n-k} \right] \tag{9.4}$$

where $P^T$is the transpose of P. Now if x is a codeword then $xH^T = 0$.

This equation can be used to determine the validity of a codeword. Let us assume c as a valid codeword and v as an arbitrary n bit word. Also assume that v represents c which incurred a nonzero error pattern e, then $v = c + e$.

Taking the product of v and H^Tgives

$$\begin{aligned} vH^T &= (c + e)H^T \\ &= cH^T + eH^T \\ &= eH^T \end{aligned} \tag{9.5}$$

So $s = vH^T$ gives the syndrome of the error. If $s = 0$ then no error occurred (except false code acceptance error).

Example 9.2 Determine the syndrome table for the (6,3) single error correcting code.

The parity-check matrix of the (6,3) code is

$$H = \begin{bmatrix} 0\,1\,1 \mid 1\,0\,0 \\ 1\,0\,1 \mid 0\,1\,0 \\ 1\,1\,0 \mid 0\,0\,1 \end{bmatrix}$$

The code has block length 5 and so we need to determine the error syndromes $s = eH^T$corresponding to the 6 single bit error patterns e = (0 0 0 0 0 1), (0 0 0 0 1 0) ... (1 0 0 0 0 0). The transpose of H is

$$H^T = \begin{bmatrix} 0\,1\,1 \\ 1\,0\,1 \\ 1\,1\,0 \\ 1\,0\,0 \\ 0\,1\,0 \\ 0\,0\,1 \end{bmatrix}$$

and as we have already seen the product eH^T gives a single row of H when e is a single bit error pattern. The error syndrome of e = (0 0 0 0 0 1) is given by the bottom row of $H^T$and so s = (0 0 1). The error syndrome of e = (0 0 0 0 1 0) is given by the second row from bottom, i.e. (0 1 0), and so forth. Table shows the resulting syndrome Table 9.2.

Table 9.2. (6, 3) Syndrome for linear block code

e	s
(0 0 0 0 0 0)	(0 0 0)
(0 0 0 0 0 1)	(0 0 1)
(0 0 0 0 1 0)	(0 1 0)
(0 0 0 1 0 0)	(1 0 0)
(0 0 1 0 0 0)	(1 1 0)
(0 1 0 0 0 0)	(1 0 1)
(1 0 0 0 0 0)	(0 1 1)

From the above syndrome table we can see that if a syndrome (0 1 0) is recognized, we can say that the error is in 5th bit and hence correct it by complementing.

9.4 Convolutional Code

This is a new form of coding technique expressed as a three-tuple (*n*, *k*, *m*) with *n* output bits for *k* input bits alongwith *m* delayed (memory) bits [2]. In the due course of study, we can visualize this technique of coding as bit by bit shifted multiplication, followed by superposition. The operation is very much similar to that of shifting, multiply through window and addition. That's why the technique is named as *convolutional coding*.

The current *n* outputs are linear combinations of the present k input bits and the previous $m \times k$ input bits.

9.4.1 Encoder

1. A binary convolutional encoder is conveniently structured as a combination of shift registers and modulo-2 adders (XOR), where the output bits are modular-2 additions of selective shift register contents and present input bits.

2. The notation n in the three-tuple notation is exactly the number of output sequences in the encoder.
3. The notation k is the number of input sequences (and hence, the encoder consists of k shift registers).
4. The value of m signifies the maximum length of the k shift registers (i.e., if the number of stages of the jth shift register is K_j, then $m = \max_{1\leq j\leq k} K_j$).
5. $K = \sum_{j=1}^{k} K_j$ is the total memory in the encoder (K is sometimes called the overall constraint lengths, popularly expressed as m+1).

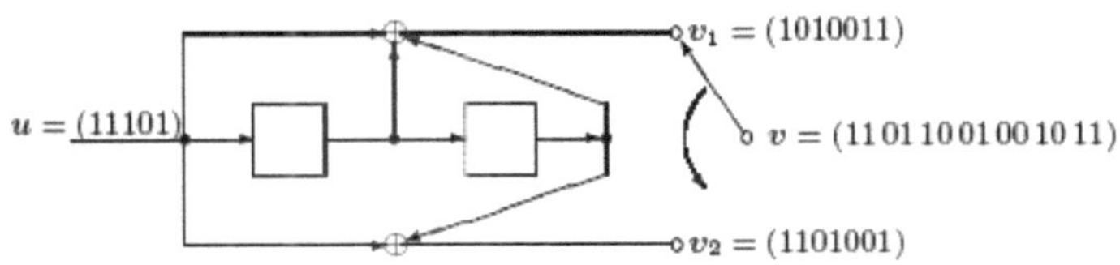

Fig. 9.3 Encoder for binary (2,1,2) convolutional code

In the above two register encoder model (Fig. 9.3), at each count of clock pulse, two bits are taken out in parallel and then converted to serial output. The outputs of the registers are fed to two different MOD-2 addition modules through two different connections.

Fig. 9.4 Encoder for binary (3,2,2) convolutional code

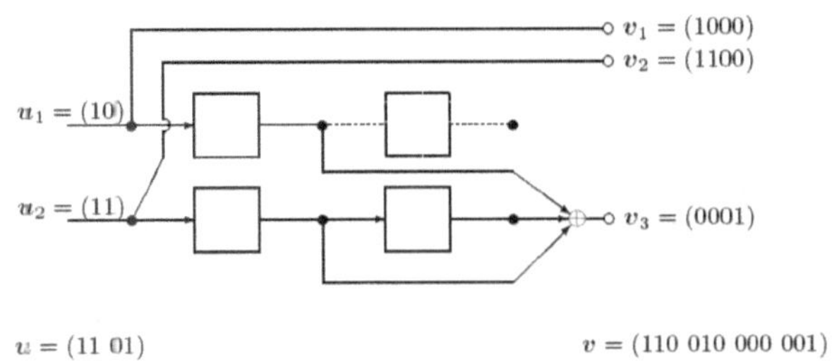

The next configuration (Fig. 9.4) takes two bits as inputs t a time and through 2 shift registers and MOD-2 addition configuration responds in two terminals in the pre-final stage. The rule of encoding must be uniform for both the cases.

9.4.1.1 Operation

Coming to the principle of operation of the convolutional encoder, we can view the entire process of encoding as a Linear Time Invariant (LTI) system. At each instance, the input bits are processed selectively through MOD-2 counter. According to Fig. 9.5, v_1 bit stream is generated as the output of the upper XOR gate. All the three inputs of that XOR gates is actively connected. Therefore, the connection matrix is designated as [1,1,1]. On the other hand, v_2 bit stream is generated as the output of the lower XOR gate. The two extreme end inputs of that XOR gates is active here. Therefore, the connection matrix is designated as [1, 0, 1].

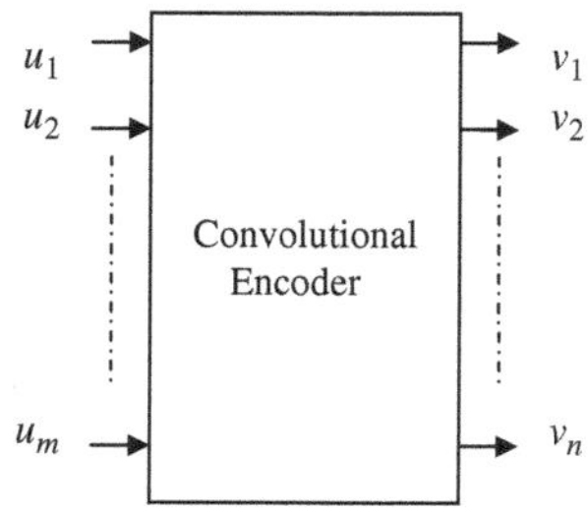

Fig. 9.5 Convolutional encoder represented as LTI system

The standard formula followed for computing convolution is

$$v_j = u_1 * h_j^{(1)} + u_2 * h_j^{(2)} + \ldots\ldots. + u_k * h_j^{(k)} = \sum_{i=1}^{k} u_i * h_j^{(i)} \tag{9.6}$$

where *is the convolution operator and $h_j^{(i)}$ is the impulse response of the ith input sequence with the response to the jth output. $h_j^{(i)}$can be easily obtained by stimulating the encoder with the discrete impulse [1,0,0,0,...] at the ith input and by observing the jth output when all the other inputs are fed the zero sequence [0,0,0,...]. Here the impulse responses are behaving like the generator polynomial or generator sequence for the convolutional encoding scheme.

From the knowledge of convolution operation, we know, there are three steps namely, *folding, shifting, multiply* and *add.* The convolutional encoder circuit suggests all those operations (the shift register is responsible for the shifting operation). Therefore, for this case of (2,1,2) convolutional code, with input bit stream $u = [1, 1, 1, 0, 1]$, two parallel impulse responses are defined as $h_1 = [1, 1, 1]$ and $h_2 = [1, 0, 1]$. The output bit stream can now be easily determined by basic rule of convolution as

$$v_1 = u * h_1 = [1, 1, 1, 0, 1] * [1, 1, 1] = [1, 0, 1, 0, 0, 1, 1] \tag{9.7}$$

$$v_2 = u * h_2 = [1, 1, 1, 0, 1] * [1, 0, 1] = [1, 1, 0, 1, 0, 0, 1] \tag{9.8}$$

The final output v is formed by taking one-by-one bit from v_1 and v_2 by a commutator. Hence, the final encoded output for this case is $v = [1, 1, 0, 1, 1, 0, 0, 1, 0, 0, 1, 0, 1, 1]$.

Similarly, for the next case of (3,2,2) convolutional code also, the impulse response can be determined as

$h_1^{(1)} = [1,0,0]$; $h_1^{(2)} = [0,0,0]$; $h_2^{(1)} = [0,0,0]$; $h_2^{(2)} = [1,0,0]$;
$h_3^{(1)} = [0,1,0]$; $h_3^{(2)} = [0,1,1]$; and similarly the final code vector obtained is: $v = [1,1,0,0,1,0,0,0,0,0,0,1]$ by the method of convolution.

9.4.1.2 Code Rate

Code rate of the convolutional encoding is defined as the ratio of message vector length to the code vector length, i.e.,

$$r = \frac{k}{n} \tag{9.9}$$

In the 1st case of (2,1,2) coding, r=1/2 and for the 2nd case of (3,2,2) coding, r=2/3.

9.4.2 State Diagram

To discuss the state diagram, code tree and trellis, let's consider the (2, 1, 2) encoder of Fig. 9.7. Previous two successive message bits (m_1 and m_2) are defined as state. The input message bit m affects the 'state' of the encoder along with the outputs v_1 and v_2. After shifting the input bit, a new state will also be generated. Let's 1st define the states of the encoder as the following Table 9.3.

Table 9.3 States of convolutional code of Fig. 9.3

m_2	m_1	State of the (2,1,2) encoder
0	0	a
0	1	b
1	0	c
1	1	d

Say, the system is in RESET position, i.e. in 'a' state. Now, the input bit is m=1. The new state and output bits v_1 and v_2 are calculated as follows (Fig. 9.6):

The output $v_1 v_2 = 11$ is transmitted for the single input bit m=1. By this way, we can draw a diagram for state transition (Fig. 9.7). Firstly, 4 circles are drawn for a=00, b=10, c=01 and d=11. 'a' to 'b' transition is done for input 1 and output 11. So, one arrow is drawn from 00 to 10 with label 1/11. Similarly, for 0 input, the

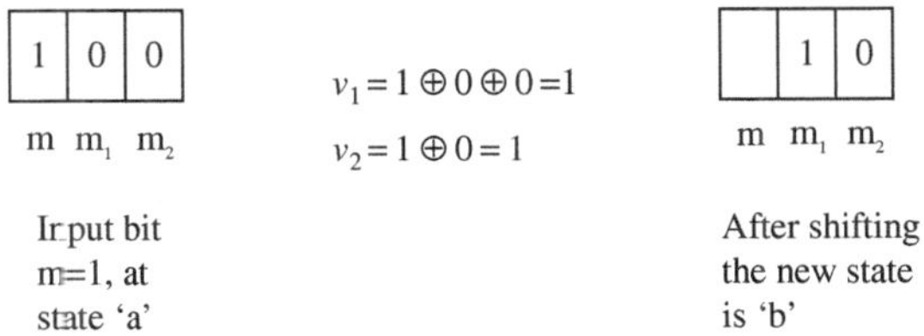

Fig. 9.6 Calculation of code vector and new state

state would not change from '00' or 'a' state. Therefore, that is a self loop with label 0/00. This procedure is followed for complete traversal of the entire state transition possibilities.

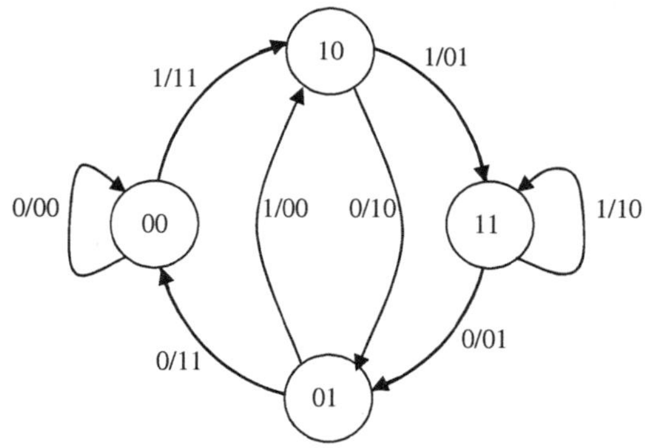

Fig. 9.7 State diagram for (2,1,2) convolutional encoder

9.4.3 Code Tree

Figure 9.8 shows the code three of the convolutional encoder. The code tree starts at node 'a' if input message bit is m=1, then path of the tree goes down towards node 'b' and output is '11'. Otherwise, if the input is m=0 at node 'a', then path of the tree goes upward towards node 'a' and output is '00'. Similarly, depending upon the input message bit, the path of the tree goes upward or downward. The nodes are marked with their states *a, b, c or d.* on the path between two nodes, the outputs are shown. We have verified the part of this code tree for 1st there message bits as '110'.

If we carefully observe the code tree of Fig. 9.8, we can find that the branch pattern begins to repeat after 3rd bit. This is shown in Fig. 9.8. The repetition starts after 3rd bit; since any message bit is stored in the shift registers of the encoder for three shifts. In the length of the shift register is increased by one bit, the pattern of code tree will repeat after 4th message bit.

9.4.4 Trellis Diagram

Trellis diagram is the more compact representation of the code tree. We know that in the code tree there are four states. Each state goes to some other state depending upon the input bit. Trellis represents the path of transition in a compact way for two types of binary inputs '0' and '1' respectively. Figure 9.9 shows the trellis diagram for (2, 1, 2) convolutional code described formerly.

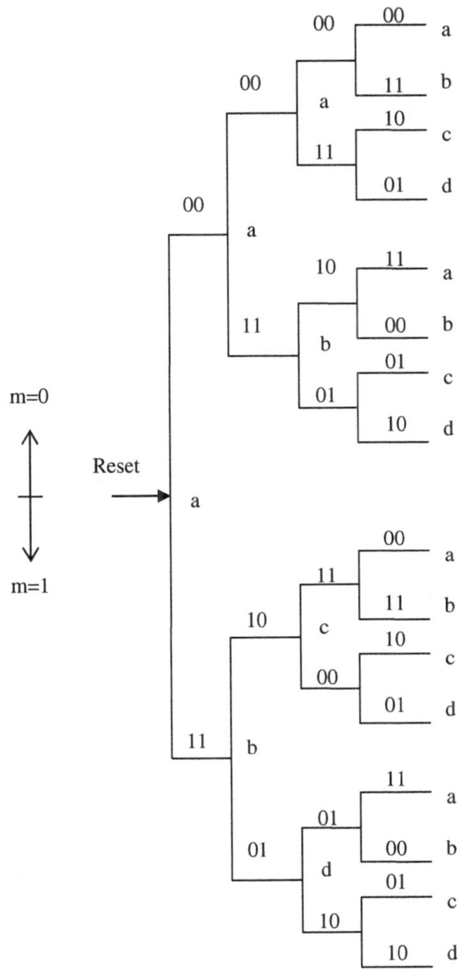

Fig. 9.8 Code tree for (2,1,2) convolutional encoder

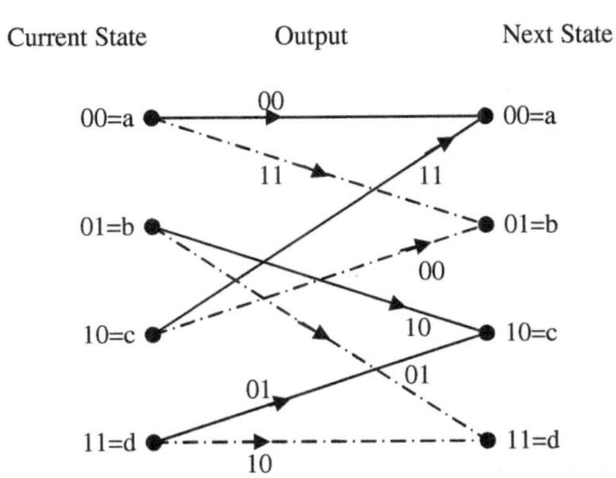

Fig. 9.9 Trellis diagram for (2,1,2) convolutional encoder

The nodes on the left denote four possible current states and those on the right represent the next state. The solid transition line represents input m=0, and hash-broken line represents input m=1. Along with each transition line the output is represented during that particular transition. As for an example, if the encoder be in state 'a' currently, for input m=0, the state will be transited to state 'a' (00) with output $v_1v_2 = 00$ and for input m=1, the state will be transited to state 'b' (01) with output $v_1v_2 = 11$. Thus the trellis diagram is the compact and convenient representation.

9.4.5 Decoding of Convolutional Code by Viterbi

Let's assume that, a coded signal is received as Y. Convolutional encoding operates continuously on input data. Hence there are non-code vectors and blocks as such. Here equiprobable (transmission error probability for symbol '1' and '0' are equal) transmission is considered. Let's define a matrix as under.

9.4.5.1 Metric

It is the discrepancy between the received signal Y and the decoded signal at a particular node. This metric may be added over few nodes for a particular path.

9.4.5.2 Surviving Path

This is the path of the decoded signal with minimum metric.

In viterbi decoding, a metric is assigned to each surviving path (i.e., metric of a particular path is obtained by adding individual metric on the nodes along the path) Y is decoded as the surviving path with smallest metric.

9.4.5.3 Principle of Decoding

Let us consider the following example of viterbi decoding. Let the signal being received is encoded by the encoder shown in the Fig. 9.3. For this encoder, we have obtained trellis of Fig. 9.9. Let the 1st six received bits be

$$Y = 11\ 01\ 11$$

It is to be noted that, for single bit input, the encoder transmits 2 bits (2, 1, 2) as output. These output bits are received at the decoder and represented by Y. Hence, Y given above represents the output bits for three successive message bits. Assume that the decoder is at state a_0.

Let us look at the trellis diagram of Fig. 9.9 for this encoder. It signifies, if the current state is 'a', then the next state will be 'a' or 'b'. This has been shown in

Fig. 9.10. Two branches are shown from a_0. One branch is at next node a_1 representing decoded symbol as '00' and other branch is at b_1 representing decoded signal as '11'.

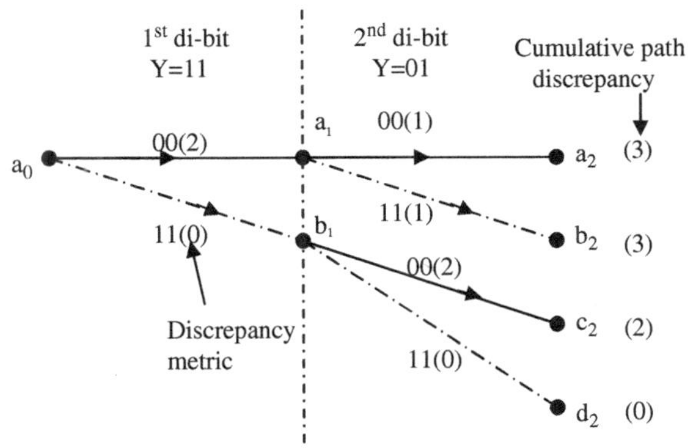

Fig. 9.10 Viterbi decoder for 1st four coded bits

The branch from a_0 to b_1 represents decoded output as '11', which is same as the 1st received duple ('11'). Hence, there is no discrepancy between the received and the decoded signal. Therefore the measured metric of that branch is zero. This metric is shown in 1st parentheses along the path of transition. When the next part of bits Y=01 is received at nodes a_1 or b_1, then from nodes a_1 and b_1 four possible next states a_2, b_2, c_2 and d_2 are possible. Figure 9.11 shows all these transitions, their decoded outputs and branch metrics corresponding to those decoded outputs. The encircled number near a_2, b_2, c_2 and d_2 show path metric emerging from a_0. Here, the path metric of path $a_0b_1d_2$ is zero.

The next figure is showing the compact diagram for the mapping of the viterbi decoding. Here s1, s2, s3, s4 are actually the states a, b, c and d with the designated levels, as signified.

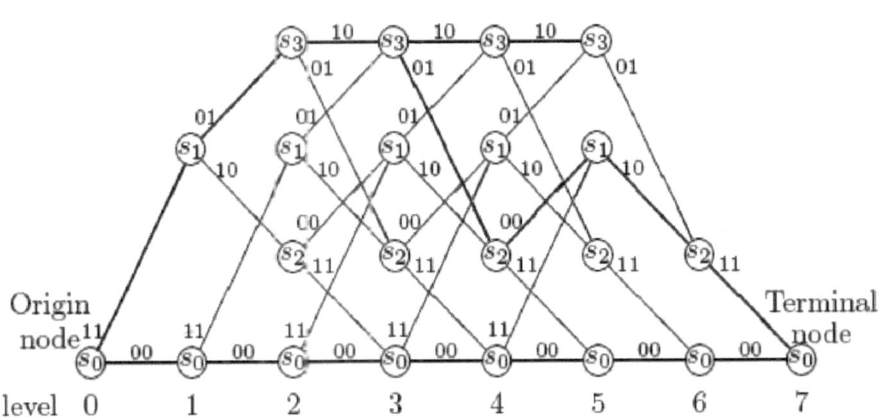

Fig. 9.11 Path mapping for viterbi decoder

9.5 Cyclic Code

9.5.1 Concept and Properties

The cyclic code may be viewed as a special class of linear block code. The mathematical structure [3] of this code permits the design of a higher order error correcting code. A code is called cyclic if a cyclic shift to the left or right yields another member of code set, considering the diagram shown in Fig. 9.1. For an example, if

$$X = (x_1 x_2 x_3 \ldots\ldots x_n) \tag{9.10}$$

is a codeword then shifting all bits cyclically to the left by one position changes the codeword from X to X^1 and

$$X^1 = (x_2 x_3 \ldots\ldots x_n x_1) \tag{9.11}$$

is also a codeword. So, in general, we can write

$$X^k = (x_{k+1} x_{k+2} x_{k+3} \ldots\ldots x_n x_1 x_2 \ldots x_k) \tag{9.12}$$

with shift in all bits to the left by k positions. Cyclic codes are easy to describe in polynomial form also. The code vector X with n bits can be expressed as a polynomial of degree (n−1) as

$$X(p) = x_1 p^{n-1} + x_2 p^{n-2} + x_3 p^{n-3} \ldots\ldots + x_n \tag{9.13}$$

where p is an arbitrary real variable. The powers of p denote the positions of the codeword bits represented by the corresponding coefficients. The coefficients of p are either 0 or 1 and follow MOD-2 addition rule. Consequently, $X(p) \oplus Y(p) = 0$ signifies that, $X(p) = Y(p)$. The code polynomial $X^k(p)$ for the generalized code vector is

$$X^k(p) = x_{k+1} p^{n-1} + x_{k+2} p^{n-2} + \ldots\ldots + x_n p^k + x_1 p^{k-1} \ldots + x_k \tag{9.14}$$

This cyclic code polynomial has an associated generator polynomial $G(p)$ so that all code words are multiple of the polynomial. The composition of $G(p)$ is related to the primitive roots of $p^n + 1 = 0$. $G(p)$ is formed by multiplying factors of the form $(p-\alpha)$, $(p-\alpha^2)$,$(p-\alpha^3)$ ..., where α is the primitive root. Some properties of the code polynomial are as follows.

1. In (n, k) cyclic code, thee exists one and only generator polynomial $G(p)$ of degree $(n-k)$ and can be expressed as

$$G(p) = p^{n-k} + g_1 p^{n-k-1} + \ldots\ldots + g_{q-1} p + p^0 \tag{9.15}$$

2. The generator polynomial $G(p)$ of an (n, k) cyclic code is a factor of $(p^n + 1)$, i.e.,

$$(p^n + 1) = G(p) \text{ [a polynomial in p of degree at most k]} \tag{9.16}$$

3. If $G(p)$ is a polynomial of degree q and is a factor of (p^n+1) then, $G(p)$ generates an (n, k) cyclic code.
4. For the message polynomial $M(p)$, the code polynomial $X(p)$ can be expressed as

$$X(p) = M(p)G(p) = \left(m_0 + m_1 p + \ldots + m_{k-1}p^{k-1}\right) G(p) \tag{9.17}$$

This method generates non-systematic code.
5. The systematic code for k message bits followed by q parity-check bits can be generated by multiplying $M(p)$ by p^q. Dividing $p^q M(p)$ by $G(p)$ we have

$$p^q M(p) = Q(p)G(p) \oplus R(p) \tag{9.18}$$

$$X(p) = p^q M(p) \oplus R(p) = Q(p)G(p) \tag{9.19}$$

where, Q(p) and R(p) are the quotient and remainder polynomials respectively.

Equation (9.18) indicates that, $p^q M(p) \oplus R(p)$ is a polynomial of degree (n−1) or less and is a multiple of $G(p)$ as well. Hence, $X(p)$ of Eq. (9.19) is a valid code vector corresponding to $M(p)$.

By simple division the property of code polynomial can be verified, i.e., if $p^q X(p)$ is divided by $p^n + 1$, it leaves a remainder $X^k(p)$ with MOD-2 operation.

Let's now consider a polynomial

$$\begin{aligned} X(p) &= M(p)G(p) \\ &= m_0 p^{k-1}G(p) + m_1 p^{k-2}G(p) + \ldots . m_{k-1}G(p) \end{aligned} \tag{9.20}$$

This is a polynomial of degree (n−1). There are a total of 2^k such polynomials corresponding to 2^k message vectors. Thus we have a linear (n, k) code generated by Eq. (9.20). To prove that the code is a cyclic code, let

$$X(p) = x_1 p^{n-1} + x_2 p^{n-2} + \ldots . x_n \tag{9.21}$$

be a polynomial in this code. Then

$$\begin{aligned} pX(p) &= x_1 p^n - x_2 p^{n-1} + \ldots . x_n p + x_1 + x_1 \\ &= x_1 \left(p^n + 1\right) + \left(x_2 p^{n-1} + \ldots . x_n p + x_1 + x_1\right) \\ &= x_1 \left(p^n + 1\right) + X^1(p) \end{aligned} \tag{9.22}$$

As $pX(p)$ is $pM(p)G(p)$ and $G(p)$ is a factor of $p^n + 1$, $X^1(p)$ must also be a multiple of $G(p)$ and can also be expressed as $M(p)G(p)$ for some message vector $M(p)$. Therefore, $X^1(p)$, $X^2(p)$ and so on are also valid code polynomials. This is the support to the statement that, the linear (n, k) code generated by $M(p)G(p)$ is a cyclic code.

9.5.2 Encoder and Decoder

Cyclic code can be of two types: systematic and non-systematic. Non-systematic $(7, 4)$ cyclic codes $X(p)$ can be easily generated from the four message bits according to Eq. (9.22). Let $M(p) = p^3 + p^2 + 1$ be the polynomial representing a message vector 1101. The generator polynomial $G(p)$ is a factor of $p^7 + 1$ and is of order n−k=7−4=3. We know, using MOD-2 addition rule,

$$p^7 + 1 = (p + 1)\left(p^3 + p^2 + 1\right)\left(p^3 + p + 1\right) \tag{9.23}$$

So either of the two the factors with order 3 can be chosen as $G(p)$ for this case. Choosing the first one, the code polynomial can be written as

$$\begin{aligned} X(p) = M(p)G(p) &= \left(p^3 + p^2 + 1\right)\left(p^3 + p^2 + 1\right) \\ &= \left(p^6 + p^4 + 1\right) \Rightarrow 1010001 \end{aligned} \tag{9.24}$$

The code polynomials for other possible combinations of four message bits are given in the Table 9.4 below.

Table 9.4 Non-systematic cyclic code for 4-bit messages and $G(p) = p^3 + p^2 + 1$

Message bits	$M(p)$	Code bits	Message bits	$M(p)$	Code bits
0001	1	0001101	0011	$p+1$	0010111
0010	p	0011010	0110	p^2+p	0101110
0100	p^2	0110100	1100	p^3+p^2	10111000
1000	p^3	1101000	0101	p^2+1	0111001
1101	p^3+p^2+1	1010001	1010	p^3+p	1110010
0111	p^2+p+1	0100011	1001	p^3+1	1100101
1110	p^3+p^2+p	1000110	1111	p^3+p^2+p+1	1001011
0000	0	0000000	1011	p^3+p+1	1111111

It is to be noted that, the cyclic shift of the code bits 1010001 constitutes the other code vectors in the left hand side of the table. Similarly, the seven code vectors at the right hand side of the table can be formed from cyclic shift of any one member in the right column. The block diagram of the encoder is shown in Fig. 9.12. Here the location of the message bits cannot be identified as such. The decoder is designed just using the complementary process.

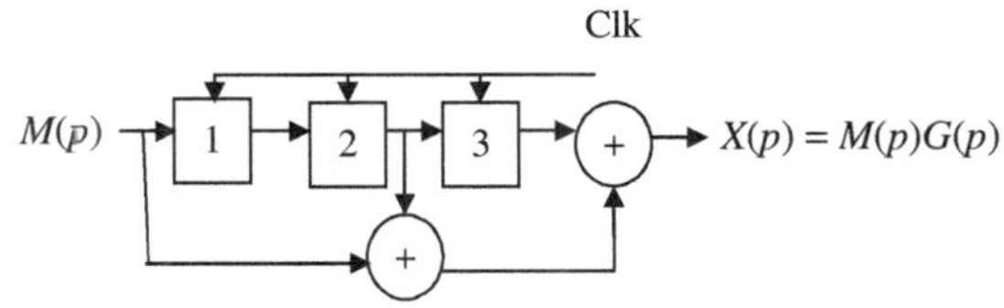

Fig. 9.12 Non-systematic cyclic encoder

If we divide $X(p)$ by $G(p)$ the quotient should be $M(p)$. Encoder for (7, 4) may be framed as follows: divide $p^3M(p)$ by $G(p)$ to obtain $R(p)$. If the received code $Y(p) = X(p)$ then in the division $Y(p) \div G(p)$ there should be no remainder. This principle is employed in the decoder.

Figures 9.13 and 9.14 shows the block diagram of the encoder and the decoder, called Meggitt decoder, based on the principle stated above. After 4 message bits input, the 3 check bits are obtained.

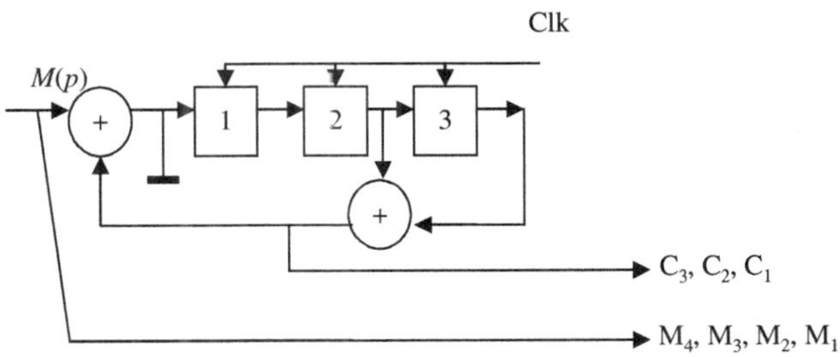

Fig. 9.13 Systematic cyclic encoder

9.5.3 Meggitt Decoder

In the Fig. 9.14, the decoder circuit is shown for (7, 4) cyclic code with $G(p) = p^3 + p^2 + 1$. All the 7 bits of the received vector Y are enter into the shift registers and the same are stored into the buffer. After the 3rd bit the process of division starts and after 7th bit, the syndrome is stored into the shift register. If the content of all the cells are {000}, Y is taken as un-erroneous. In Meggitt decoder a typical pattern of the syndrome {100} appears whenever the erroneous bit is shifted out of buffer. This is used through a coincidence circuit to correct the erroneous circuit as shown in the Fig. 9.14.

9.6 BCH Code

BCH codes were invented in 1959 by Hocquenghem, and independently in 1960 by Bose and Ray-Chaudhuri [6]. The acronym *BCH* comprises the initials of these inventors' names. This is a multilevel cyclic variable-length error-correcting code used to correct multiple random error patterns. BCH codes may also be used with

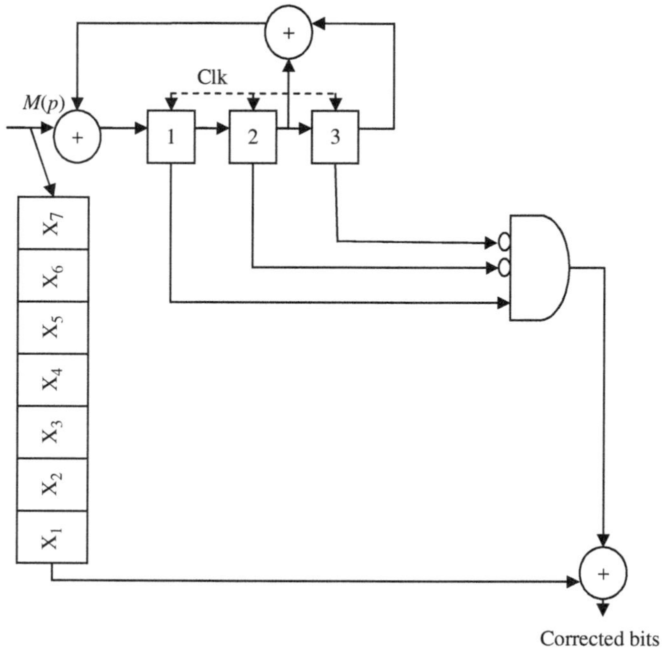

Fig. 9.14 Meggitt decoder

multilevel phase-shift keying whenever the number of levels is a prime number or a power of a prime number. A BCH code in 11 levels has been used to represent the 10 decimal digits plus a sign digit.

9.6.1 Simplified BCH Codes

Definition 9.1 *Fix a finite field* $GF(q^m)$, *where* q *is a prime. Also fix positive integers* n, *and* d *such that* $n = q^m - 1$ *and* $2 \leq d \leq n$. *We will construct a polynomial code over* GF(q) *with code length* n *(generator polynomial), whose minimum Hamming distance is at least* d.

Let α be a primitive nth root of unity in $GF(q^m)$. For all *i*, 14t $m_i(x)$ be the minimal polynomial of α^i with coefficients in $GF(q)$. The generator polynomial of the BCH code is defined as the least common multiple

$$g(x) = lcm(m_1(x), m_2(x), \ldots m_{d-1}(x)) \tag{9.25}$$

With a simple example we can try to understand the technique.

Let $q = 2$ and $m = 4$ (therefore $n = 15$). We will consider different values of *d*. There is a primitive root $\alpha \in GF(16)$ satisfying the following equation.

$$\alpha^4 + \alpha + 1 = 0 \tag{9.26}$$

its minimal polynomial over $GF(2)$ is: $m_1(x) = x^4 + x + 1$. Note that in $GF(2^4)$, the equation $(a + b)^2 = a^2 + 2ab + b^2 = a^2 + b^2$ holds, and therefore $m_1(\alpha^2) = m_1(\alpha)^2 = 0$. Thus α^2 is a root of $m_1(x)$, and therefore

$$m_2(x) = m_1(x) = x^4 + x + 1. \tag{9.27}$$

To compute $m_3(x)$, we have the following linear relations:

- $1 = 0\alpha^3 + 0\alpha^2 + 0\alpha + 1$
- $\alpha^3 = 1\alpha^3 + 0\alpha^2 + 0\alpha + 0$
- $\alpha^6 = 1\alpha^3 + 1\alpha^2 + 0\alpha + 0$
- $\alpha^9 = 1\alpha^3 + 0\alpha^2 + 1\alpha + 0$
- $\alpha^{12} = 1\alpha^3 + 1\alpha^2 + 1\alpha + 1$

Five right-hand-sides of length four must be linearly dependent, and indeed we find a linear dependency $\alpha^{12} + \alpha^9 + \alpha^6 + \alpha^3 + 1 = 0$. Since there is no smaller degree dependency, the minimal polynomial of α^3 is: $m_3(x) = x^4 + x^3 + x^2 + x + 1$. Continuing in a similar manner, we find

$$\begin{aligned} m_4(x) &= m_2(x) = x^4 + x + 1, \\ m_5(x) &= x^2 + x + 1, \\ m_6(x) &= m_3(x) = x^4 + x^3 + x^2 + x + 1, \\ m_7(x) &= x^4 + x^3 + 1 \end{aligned} \tag{9.28}$$

The BCH code with d = 1,2,3 has generator polynomial

$$d(x) = m_1(x) = x^4 + x + 1.$$

It has minimal Hamming distance at least 3 and corrects up to 1 error. Since the generator polynomial is of degree 4, this code has 11 data bits and 4 checksum bits.

The BCH code with d = 4,5 has generator polynomial

$$\begin{aligned} d(x) &= \mathrm{lcm}\,(m_1(x), m_3(x)) \\ &= \left(x^4 + x + 1\right)\left(1 + x + x^2 + x^3 + x^4\right) = x^8 + x^7 + x^6 + x^4 + 1 \end{aligned} \tag{9.29}$$

It has minimal Hamming distance at least 5 and corrects up to 2 errors. Since the generator polynomial is of degree 8, this code has 7 data bits and 8 checksum bits.

9.6.2 General BCH Codes

General BCH codes differ from the simplified case discussed above in two respects. First, one replaces the requirement $n = q^m - 1$ by a more general condition. Second, the consecutive roots of the generator polynomial may run from $\alpha^c, \ldots, \alpha^{c+d-2}$ instead of $\alpha, \ldots, \alpha^{d-1}$.

Definition 9.2 *Fix a finite field* GF(q), *where* q *is a prime power. Choose positive integers* m,n,d,c *such that* $2 \leq d \leq n$, *gcd*(n,q) = 1, *and* m *is the multiplicative order of* q *modulo* n.

As before, let α be a primitive nth root of unity in $GF(q^m)$, and let $m_i(x)$ be the minimal polynomial over $GF(q)$ of α^i for all i. The generator polynomial of the BCH code is defined as the least common multiple

$$g(x) = lcm\,(m_c(x), m_{c+1}(x), \ldots .m_{c+d-2}(x)) \tag{9.30}$$

9.6.3 Properties

1. The generator polynomial of a BCH code has degree at most $(d-1)\,m$. Moreover, if $q = 2$ and $c = 1$, the generator polynomial has degree at most $dm/2$.
   Proof Each minimal polynomial $m_i(x)$ has degree at most m. Therefore, the least common multiple of $d - 1$ of them has degree at most $(d - 1)m$. Moreover, if $q = 2$, then $m_i(x) = m_{2i}(x)$ for all i. Therefore, $g(x)$ is the least common multiple of at most $d/2$ minimal polynomials $m_i(x)$ for odd indices i, each of degree at most m.
2. A BCH code has minimal Hamming distance at least d.
   Proof We only give the proof in the simplified case; the general case is similar. Suppose that $p(x)$ is a code word with fewer than d non-zero terms. Then

$$p(x) = b_1 x^{k_1} + \ldots. + b_{d-1} x^{k_{d-1}} \tag{9.31}$$

   Recall that $\alpha^1 \ldots, \alpha^{d-1}$ are roots of $g(x)$, hence of $p(x)$. This implies that $b_1, \ldots .b_{d-1}$ satisfy the following equations, for $i = 1, \ldots, d-1$.

$$p\left(\alpha^i\right) = b_1 \alpha^{ik_1} + b_2 \alpha^{ik_2} \ldots\ldots + b_{d-1} \alpha^{ik_{d-1}} = 0 \tag{9.32}$$

   In matrix form, we have

$$\begin{bmatrix} \alpha^{k_1} & \alpha^{k_2} & \cdots & \alpha^{k_{d-1}} \\ \alpha^{2k_1} & \alpha^{2k_2} & \cdots & \alpha^{2k_{d-1}} \\ \vdots & & & \\ \alpha^{(d-1)k_1} & \alpha^{(d-1)k_2} & \cdots & \alpha^{(d-1)k_{d-1}} \end{bmatrix} \begin{bmatrix} b_1 \\ b_2 \\ \vdots \\ b_{d-1} \end{bmatrix} = \begin{bmatrix} 0 \\ 0 \\ \vdots \\ 0 \end{bmatrix} \tag{9.33}$$

3. A BCH code is cyclic.

Proof A polynomial code of length n is cyclic if and only if its generator polynomial divides $x^n - 1$. Since $g(x)$ is the minimal polynomial with roots $\alpha^c, \ldots, \alpha^{c+d-2}$, it suffices to check that each of $\alpha^c, \ldots, \alpha^{c+d-2}$ is a root of $x^n - 1$. This follows immediately from the fact that α is, by definition, an nth root of unity.

References

1. Carlson, A.B., Crilly, P.B., Rutledge, J.C., "Communication systems – An Introduction to Signals and Noise in Electrical Communication (4th Ed.)", McGraw-Hill International, 2002
2. Bossert, M., "Channel Coding for Telecommunications". New York, NY: John Wiley and Sons, 1999
3. Sharma, S., "Communication Systems (Analog and Digital) (4th Ed.)", S.K. Kataria and Sons, 2007
4. Meggitt, J.E., "Error Correcting Codes and their Implementation", IRE Transactions on Information Theory, Vol. IT-7, 1961, pp. 232–244
5. Hocquenghem, A., "Codes correcteurs d'erreurs", Chiffres (Paris), Vol. 2, September 1959, pp. 147–156
6. Bose, R.C., Ray-Chaudhuri, D.K., "On a Class of Error Correcting Binary Group Codes", Information and Control, Vol. 3, No. 1, March 1960, pp. 68–79

Appendix A
Elementary Probability Theory

A.1 Concept of Probability

In communication system, prediction of system performance, noise accumulation probability, probability distribution of noise are really essential studies. At a particular instant, one physical *experiment* may have a deterministic outcome or indeterministic outcome. In this appendix, the mathematical models of experiments are prepared. A single performance of the physical experiment is called as *trial* for which there is an *outcome*.

A.1.1 Random Experiments and Sample Space

The experiments, in which results are not being essentially the same even though conditions may be nearly identical, are called *random experiments*.

The following are some examples.

(a) *Tossing a coin*: The result of the experiment is that it will either come up 'tails,' symbolized by T (or 0), or 'heads,' symbolized by H (or 1), i.e. one of the elements of the set {H, T} (or {0, 1}).
(b) *Tossing a die*: The result of the experiment is that it will come up with one of the numbers in the set {1, 2, 3, 4, 5, 6}.
(c) *Tossing a coin twice*: The results can be indicated by {HH, HT, TH, TT}, i.e. both heads, head on first and tail on second, etc.

A set S which consists of all possible outcomes of a random experiment is called a *sample space* and each outcome is called a *sample point* [1].

To understand sample space, let's take the example of tossing a die, one sample space or set of all possible outcomes is given by {1, 2, 3, 4, 5, 6} while another is {odd, even}.

If the sample space S is a set of finite number of elements, then the space is *discrete* sample space like tossing a die or a coin. There are sample spaces where

A. Das, *Digital Communication*, Signals and Communication Technology,
DOI 10.1007/978-3-642-12743-4, © Springer-Verlag Berlin Heidelberg 2010

number of outcome is infinite/uncountable. For those cases, the sample space is defined as *continuous.*

A.1.2 Events

An event is a subset A of the sample space S, i.e. it is a set of possible outcomes. If the outcome of an experiment is an element of A we say that the event A has occurred. An event consisting of a single point of S is often called a simple or *elementary event.*

As an example, let's take the random experiment of tossing a coin twice. Here, two events are defined as A and B, where A be the event that 'at least one head occurs' and B be the event that 'the second toss results in a tail.' Then A = {HT, TH, HH}, B = {HT, TT},

A.1.3 Probability-Understanding Approaches

From the study of random experiment, it is clear that, in any random experiment there is always uncertainty as to whether a particular event will or will not occur. As a measure of the chance or probability with which we can expect the event to occur it is convenient to assign a number between 0 and 1. If we are sure or certain that the event will occur we say that its probability is 100% or 1, but if we are sure that the event will not occur we say that its probability is zero.

Definition approach 1 If an event can occur in *h* different ways out of a total number of *n* possible ways, all of which are equally likely, then the probability of the event is *h*/*n* [1].

Definition approach 2 If after *n* (*n* is a large number) repetitions of an experiment, an event is observed to occur in *h* of these, then the probability of the event is *h*/*n*. This is also called the *empirical probability* of the event.

Mathematically, probability of a favorable event h from a sample space S is defined as

$$p(h) = \underset{n \to \infty}{Lt} \frac{h}{n} \tag{A.1}$$

where, *n* is the total number of chance experiments.

A.2 Random Variable

Random variable is neither *random* nor *variable* [2]; conversely, it can be defined as a *function* of the elements of a sample space *S*. Generally random variable is represented by a capital letter (X) and any particular value of the random variable by a lower case letter (x). Thus, given an experiment defined by a sample space S with elements s, we assign to every s a real number $X(s)$ according to some unique mapping and $X(s)$ is called as random variable.

The random variable considered being a function or mapping that maps all the elements of the sample space into points on the real line. The following example can make our understanding clear.

Figure A.1 shows an experiment where the pointer on a wheel of chance is spun. The possible outcomes are the numbers from 0 to 12 marked on the wheel. The sample space consists of the numbers is the set $\{0 < s \leq 12\}$. We define a random variable by the function

$$X = X(s) = s^2 \tag{A.2}$$

Points in S now map onto the real line as the set $\{0 < s \leq 144\}$.

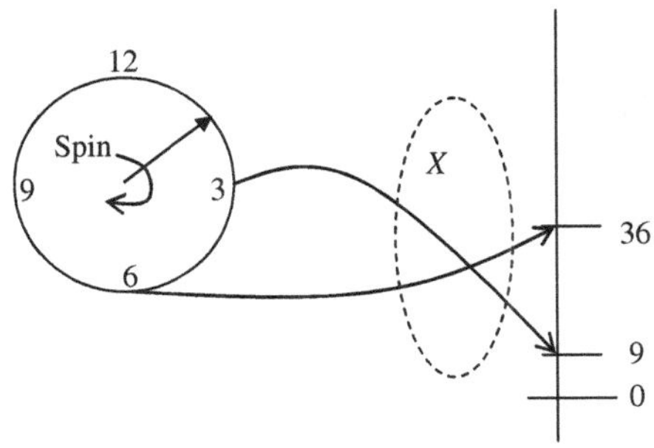

Fig. A.1 Random variable as a mapping/function of a particular experiment

A.3 Mean, Variance, Skew-ness and Kurtosis

The *mean* [3] of a data set is simply the arithmetic average of the values in the set, obtained by summing the values and dividing by the number of values. Say, the events $x_1, x_2, x_3 \ldots x_N$ are occurring with frequencies $n_1, n_2, n_3 \ldots n_N$ respectively. Therefore the average of occurrence of the events is given by

$$\begin{aligned}
\text{Mean} = \mu &= \frac{x_1 n_1 + x_2 n_2 + \ldots\ldots + x_N n_N}{n_1 + n_2 + \ldots\ldots + n_N} = \frac{x_1 n_1 + x_2 n_2 + \ldots\ldots + x_N n_N}{N} \\
&= x_1 \frac{n_1}{N} + x_2 \frac{n_2}{N} + \ldots\ldots + x_N \frac{n_N}{N} = x_1 p(x_1) + x_2 p(x_2) + \ldots\ldots + x_N \\
\therefore \mu &= \sum_{i=1}^{N} x_i p(x_i)
\end{aligned} \tag{A.3}$$

The *variance* of a data set is the arithmetic average of the squared differences between the values and the mean. Again, when we summarize a data set in a frequency distribution, we are approximating the data set by 'rounding' each value in a given class to the class mark. Thus, the variance of a frequency distribution is given by

$$\sigma^2 = \sum_{i=1}^{N} (x_i - \mu)^2 p(x_i) \tag{A.4}$$

The *standard deviation* is the square root of the variance.

Skew-ness [3] is the third order moment around the mean. If two distributions are given as shown in the Fig. A.2, the distribution of Fig. A.2b is more skewed than Fig. A.2a. Skew-ness signifies the asymmetry in the distribution.

$$\text{Skew-ness } \zeta = \sum_{i=1}^{N} (x_i - \mu)^3 p(x_i) \tag{A.5}$$

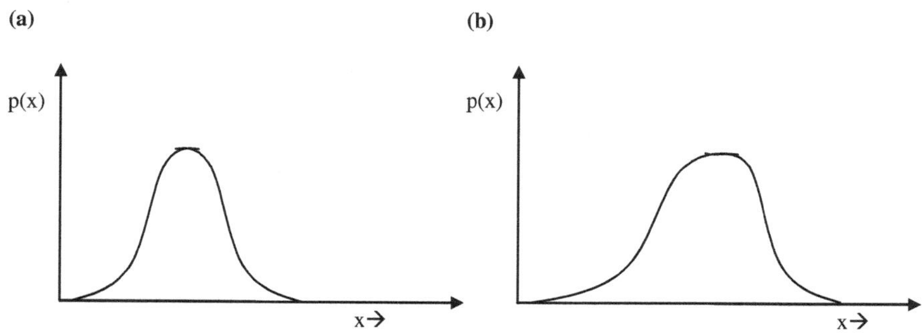

Fig. A.2 Physical interpretation of skew-ness

Kurtosis is the fourth order moment of the distribution around the mean. If two distributions are given as shown in the Fig. A.3, the distribution of Fig. A.3b is of

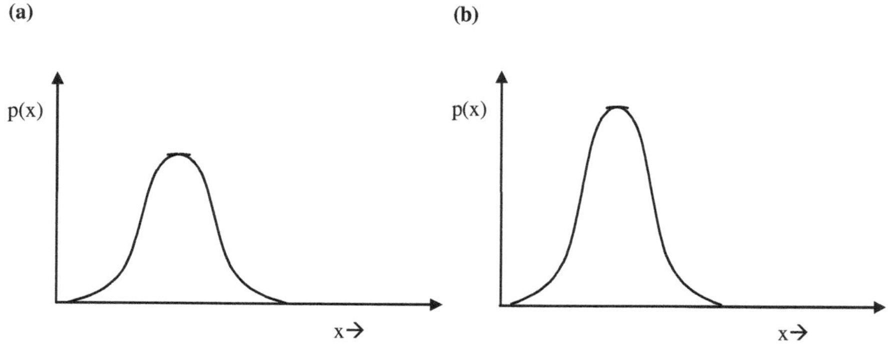

Fig. A.3 Physical interpretation of kurtosis

more value of coefficient of kurtosis than that of Fig. A.3a. Kurtosis signifies the pickiness in the distribution.

$$\text{Kurtosis } \alpha^4 = \sum_{i=1}^{N} (x_i - \mu)^4 \, p(x_i) \tag{A.6}$$

A.4 Cumulative Distribution Function (CDF)

Cumulative distribution function (CDF) associated with a random variable is defined as the probability that the outcome of an experiment will be one of the outcomes for which $X(\lambda) \leq x$ [2].

$$F_X(x) \triangleq p(X \leq x)\,; \; -\infty \leq x \leq \infty \tag{A.7}$$

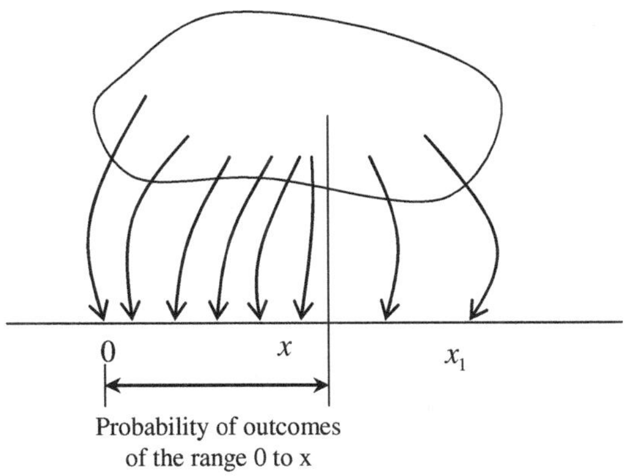

Fig. A.4 Cumulative distribution function (CDF)

Example A.1 Tossing two dies
The outcomes are

$$\begin{array}{ll}
(1,1),(1,2),(1,3),\ldots\ldots\ldots(1,6) & \Rightarrow 6 \\
(2,1),(2,2),(2,3),\ldots\ldots\ldots(2,6) & \Rightarrow 6 \\
(3,1),(3,2),(3,3),\ldots\ldots\ldots(3,6) & \Rightarrow 6 \\
\vdots & \\
\vdots & \\
(6,1),(6,2),(6,3),\ldots\ldots\ldots(6,6) & \Rightarrow 6 \\
 & \overline{\quad 36 \quad}
\end{array}$$

Let us assume that, $\lambda_{i,j}$ is a number which can take 36 values (possible variables), where $1 \leq i \leq 6$ and $1 \leq j \leq 6$.
Now, the random variable(X) is defined such that,
$X\left(\lambda_{i,j}\right) = i + j = n$; where $2 \leq n \leq 12$.
This 'n' is a random number defined by the above relationship.

$\underline{p(n):}$

$$p(0) = p(1) = 0 \qquad p(2) = \frac{1}{36}$$

$$p(3) = \frac{2}{36} \qquad p(4) = \frac{3}{36}$$

$$p(5) = \frac{4}{36} \qquad p(6) = \frac{5}{36}$$

$$p(7) = \frac{6}{36} \qquad p(8) = \frac{5}{36} = p(6)$$

$$p(9) = \frac{4}{36} = p(5) \qquad p(10) = \frac{3}{36} = p(4)$$

$$p(11) = \frac{2}{36} = p(3) \qquad p(12) = \frac{1}{36} = p(2)$$

Therefore, the CDF can be determined as

$$F_X(x) \triangleq p(X \leq n)$$

$$F(0) = p(0) = 0 = F(1); \qquad F(2) = p(0) + p(1) + p(2) = \frac{1}{36}$$

$$F(3) = F(2) + p(3) = \frac{3}{36}; \qquad F(4) = F(3) + p(4) = \frac{6}{36}$$

$$F(5) = F(4) + p(5) = \frac{10}{36}; \qquad F(6) = F(5) + p(6) = \frac{15}{36}$$

$$F(7) = \frac{21}{36}; \qquad F(8) = \frac{26}{36}$$

$$F(9) = \frac{30}{36}; \qquad F(10) = \frac{33}{36}$$

$$F(11) = \frac{35}{36}; \qquad F(12) = \frac{36}{36} = 1$$

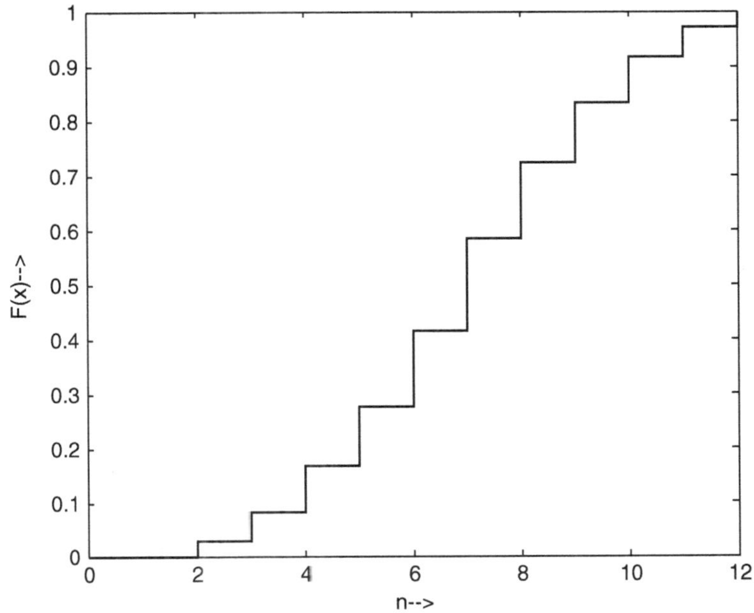

Fig. A.5 Plot of CDF with respect to a random variable

A.5 Probability Density Function (PDF)

Probability Density Function (PDF) associated with a random variable is defined as the slope of the cumulative distribution function at a particular point of random variable $X(\lambda) \leq x$.

$$f_X(x) \triangleq \frac{d}{dx}[F_X(x)] \tag{A.8}$$

From the above example, PDF can also be plotted as the Fig. A.6 below.

A.5.1 Uniform PDF

Uniform PDF is defined as the PDF which is constant over a certain range (Fig. A.7). We can try to understand the concept using an example given below.

$$\begin{aligned} f_X(x) &= c; 0 \leq x \leq 2\pi \\ &= 0 \quad \text{otherwise} \end{aligned} \tag{A.9}$$

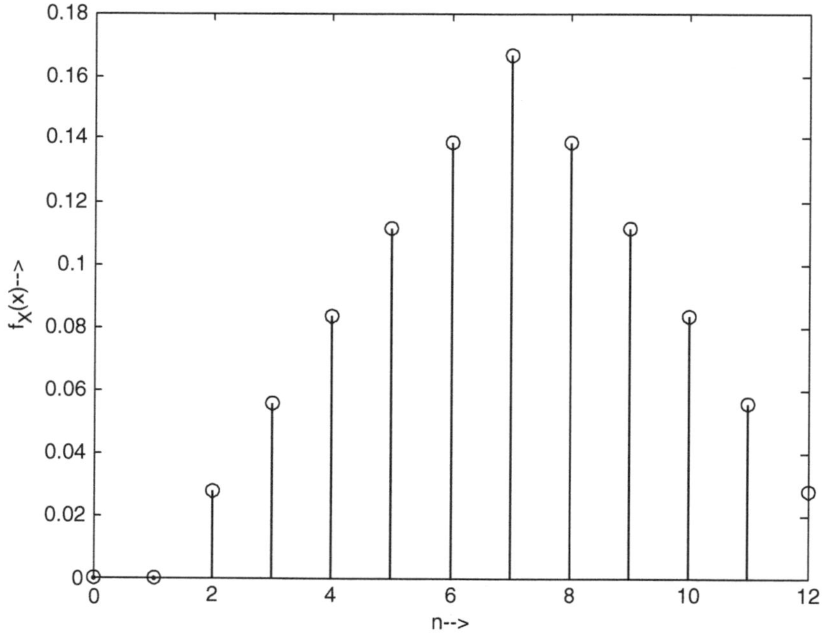

Fig. A.6 Plot of PDF with respect to a random variable

Fig. A.7 Uniform PDF

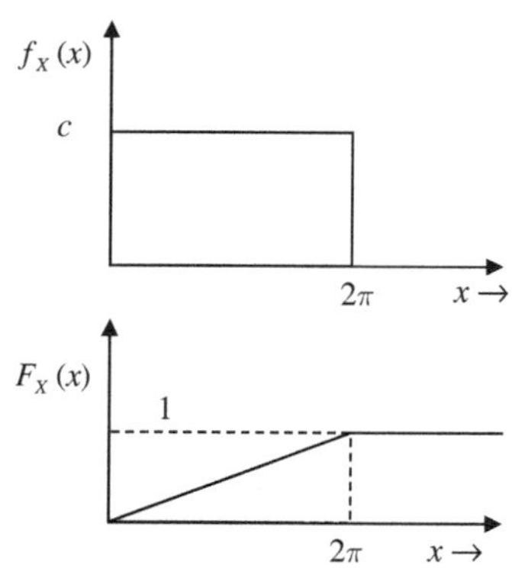

A.5.2 Frequently Used Probability Distribution

A.5.2.1. Bernoulli Distribution

$$p(x) =^n C_x p^x (1 - p)^{n-x} \tag{A.10}$$

In the Fig. A.8, the Bernoulli distribution is shown as a discrete distribution having two possible outcomes for x=0 and x=1 with respective probabilities 0.6 and 0.4.

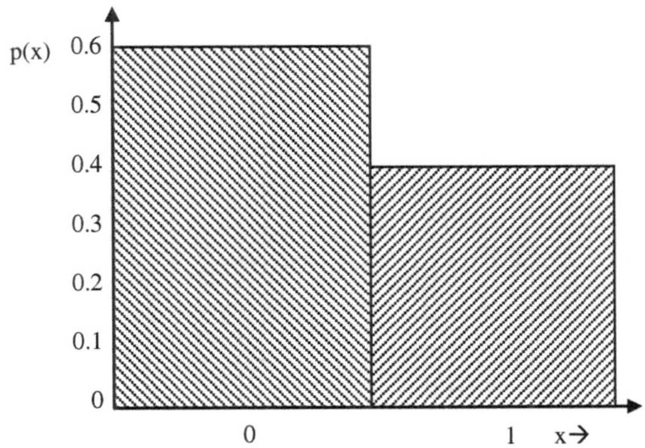

Fig. A.8 Bernoulli distribution for p=0.4

$$p(x) = \begin{cases} p. & for \quad x = 1 \\ (1-p) & for \quad x = 0 \end{cases} \tag{A.11}$$

A.5.2.2 Gaussian Distribution

$$p(x) = \frac{1}{\sqrt{2\pi\sigma^2}} e^{-(x-\mu)^2/2\sigma^2} \tag{A.12}$$

where, μ is the mean and σ is the standard deviation.

2-D Gaussian can also be defined as

$$p(x, y) = \frac{1}{\sqrt{2\pi\sigma^2}} e^{-\{(x-\mu)^2/2\sigma^2+(y-\mu)^2/2\sigma^2\}} \tag{A.13}$$

A.5.2.3. Poisson Distribution

$$p(x, \lambda) = \frac{\lambda^x e^{-\lambda}}{x!}; \lambda = 0, 1, 2 \ldots\ldots \tag{A.14}$$

where λ is the occurrence interval [4].

A.5.2.4 Rayleigh Distribution

$$p(x) = \frac{x}{\sigma^2} \exp\left(-\frac{x^2}{2\sigma^2}\right) \tag{A.15}$$

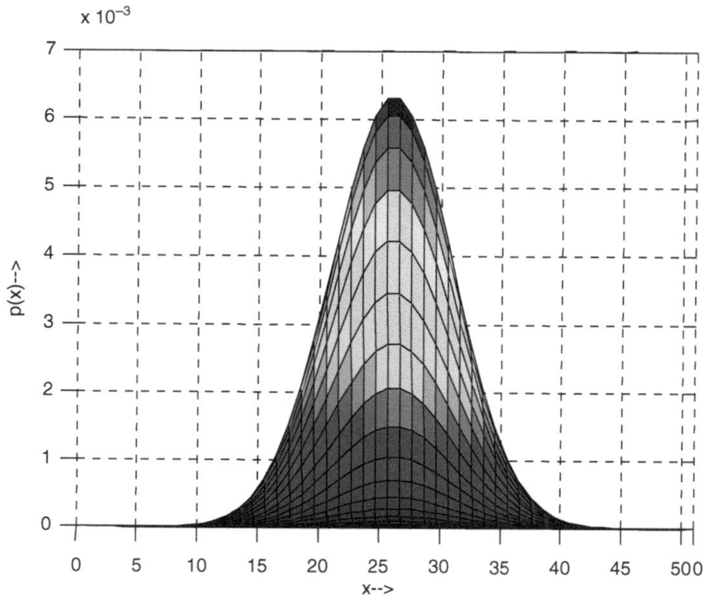

Fig. A.9 1-D Gaussian distribution

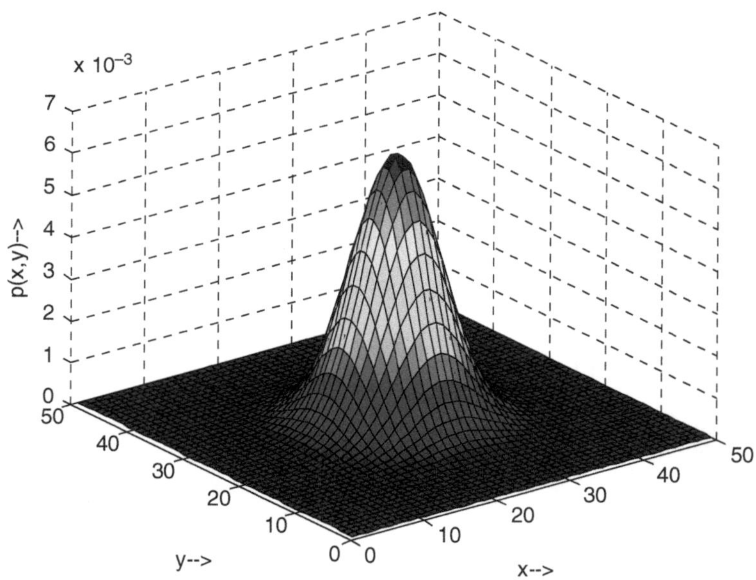

Fig. A.10 2-D Gaussian distribution

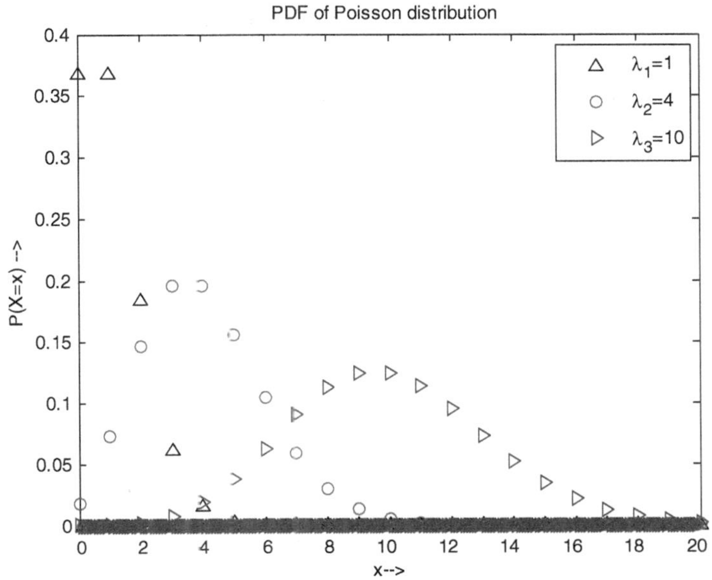

Fig. A.11 Probability mass function for Poisson distribution

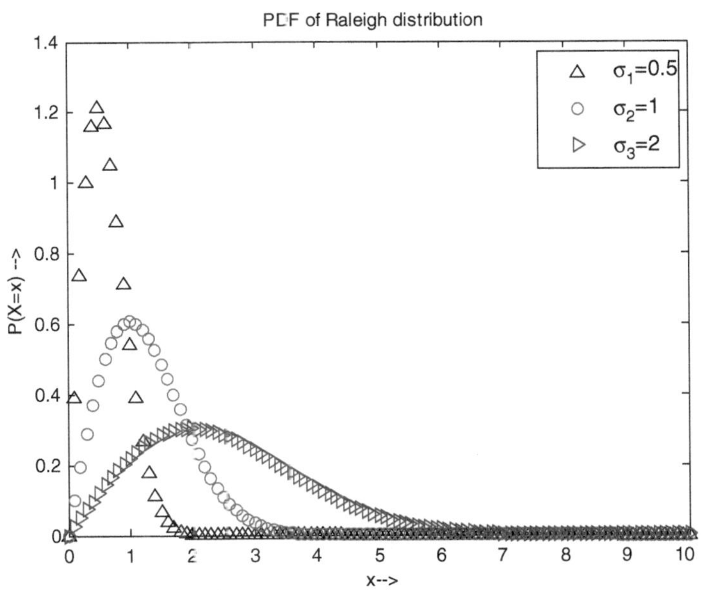

Fig. A.12 Probability mass function for Raleigh distribution

References

1. Peyton Z. Peebles, Jr., "Probability, Random Variables, and Random signal Principles (2nd Ed.,)", McGraw-Hill, 1987
2. Carlson, A.B., Crilly, P.B., Rutledge, J.C., "Communication Systems – An Introduction to Signals and Noise in Electrical Communication (4th Ed.)", McGraw-Hill International, 2002
3. Spiegel, M.R., "Schaum's Outline of Theory and Problems of Probability and Statistics", McGraw-Hill, 1998
4. http://mathworld.wolfram.com

Appendix B
Convolution and Correlation – Some Case Studies

B.1 Convolution

Generally, we define a system (like a digital communication system) as a function or mapping which maps the input sequence (sampled signal) to output sequence. The mapping rule may be defined clearly or not. Say, the output or response $y(n)$ of a system τ with input or excitation $x(n)$ is clearly defined as $y(n) = 0.5x^2(n)$. If that is the case, then we can predict the output for any query input, accurately. But in practice, the situation is not so easy. The mapping τ is unknown in most of the cases and we have to consider the system as a black box as shown in the Fig. B.1.

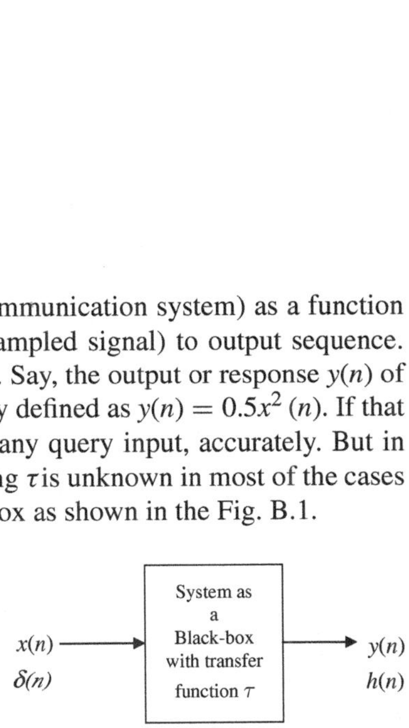

Fig. B.1 Digital system with unknown transfer function

Now, if one asks for the predicted output for any input, there must also be some procedure taking the help of the black-box. With this objective let us try to define any given sequence $x(n)$ as Eq. (B.1).

$$\begin{aligned} x(n) &= x(-\infty)\delta(n+\infty) + \ldots\ldots\ldots\ldots\ldots + x(-1)\delta(n+1) + x(0)\delta(n) + \\ &\quad x(1)\delta(n-1) + \ldots x(\infty)\delta(n-\infty) \\ \Rightarrow x(n) &= \sum_{k=-\infty}^{\infty} x(k)\delta(n-k) \end{aligned} \tag{B.1}$$

This is true for any sequence of infinite length, for finite length sequence the limits of summation of Eq. (B.1) will be finite. We know that $\delta(n)$ is the unit impulse defined as

$$\delta(n) = \begin{cases} 0; n \neq 0 \\ 1; n = 0 \end{cases} \tag{B.2}$$

And the impulse response of the system is defined as the output (response) of the present system with input (excitation) unit impulse, given by

$$h(n) = \dot{\tau}\,[\delta(n)] \tag{B.3}$$

Now, if this sequence is feed into the system τ, the output must be given by

$$\begin{aligned} y(n) &= \zeta\,[x(n)] \\ &= \zeta\left[\sum_{k=-\infty}^{\infty} x(k)\delta(n-k)\right] \\ &= \sum_{k=-\infty}^{\infty} x(k)\zeta\,[\delta(n-k)] \qquad \text{[If the system is Linear]} \\ &= \sum_{k=-\infty}^{\infty} x(k)h(n-k) \qquad \text{[If the system is Time Invariant]} \\ &\Rightarrow x(n)^{*}h(n) \end{aligned} \tag{B.4}$$

The operation given by Eq. (B.4) is the convolution. This is not similar to simple multiplication. Here four operations are involved namely *folding, multiply, add, shifting*. The operations are understandable from the Eq. (B.4).

First the impulse response is obtained from the system (black-box) with the unit impulse excitation. Next, the impulse response is folded to get $h(-k)$ with n=0 and taking k as the index. Then all the samples are multiplied with $x(k)$ in sample by sample basis and added up to get $y(0)$. To get other output samples $y\,(n \neq 0)$, $h(-k)$ is to be shifted to both of the sides of time axis or sequence count (n) axis. The operations are described using an example as shown in Fig. B.2. Say the impulse response $h(n) = \{1, -1\}$ and the input sequence is $x(n) = \{-1, -1, -1, 1, 1, 1\}$.

In the example shown in Fig. B.2, $y(n)$ is 0 except $n = 2$. The physical interpretation of this example is very important to understand any component of digital communication systems. Here, the impulse response $h(n)$ itself is behaving as the transfer function of the system. We can notice that the impulse response is actually combination of a +1 followed by a -1. The combination results in a derivative, as in digital domain differential equation become difference equation. If any input having that differential property is feed into that particular system, the system $(h(n))$ will make the property prominent at the output. Therefore, the system characteristic suggests the system $h(n)$ to be a *differentiator.* Similarly, the impulse response can be of filter type, of modulator type and so on. The convolution operation is applicable for linear time invariant (LTI) systems only.

Some basic properties with necessary proofs are discussed in the next subsection.

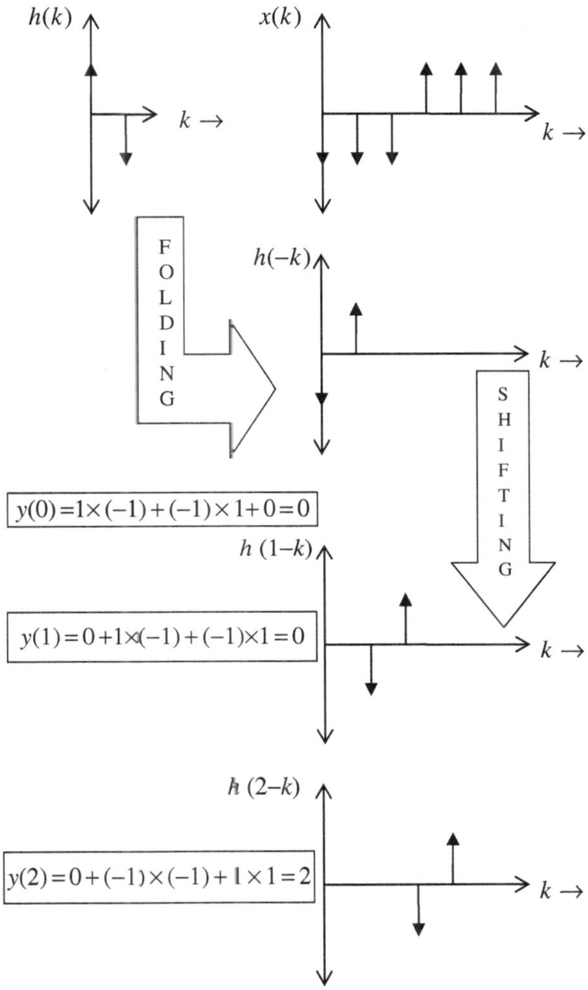

Fig. B.2 Convolution operation between two sequences

B.1.1 Basic Properties of Convolution

B.1.1.1 Commutative Law

$$x_1(t)^* x_2(t) = x_2(t)^* x_1(t) \tag{B.5}$$

Proof From the basic definition of convolution in discrete domain referring Eq.(B.4), we can also express convolution in continuous domain as

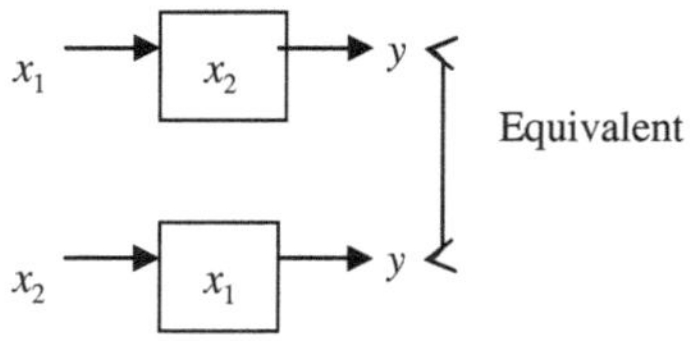

Fig. B.3 Commutative law: either function can be regarded as the input while the other as impulse response of the system

$$x_1(t)^* x_2(t) = \int_{-\infty}^{\infty} x_1(\tau) x_2(t-\tau)\, d\tau \tag{B.6}$$

Let, $(t-\tau) = k$, therefore, $\tau = (t-k)$. Therefore, the Eq.B.6 can be re-written as

$$\begin{aligned} x_1(t)^* x_2(t) &= \int_{-\infty}^{\infty} x_1(\tau) x_2(t-\tau)\, d\tau \\ &= \int_{-\infty}^{\infty} x_2(k) x_1(t-k)\, dk \\ &= x_2(t)^* x_1(t) \end{aligned} \tag{B.7}$$

B.1.1.2 Associative Law

$$x_1(t)^* \left(x_2(t)^* x_3(t)\right) = \left(x_1(t)^* x_2(t)\right)^* x_3(t) \tag{B.8}$$

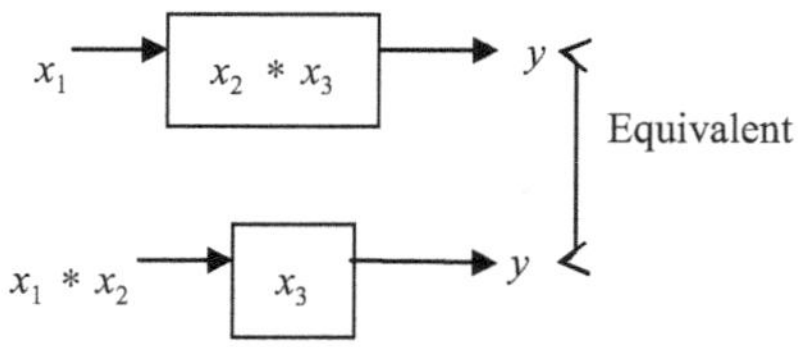

Fig. B.4 Associative property of convolution

B.1.1.3 Distributive Law

$$x_1(t)^* \left(x_2(t) + x_3(t)\right) = \left(x_1(t)^* x_2(t)\right) + \left(x_1(t)^* x_3(t)\right) \tag{B.9}$$

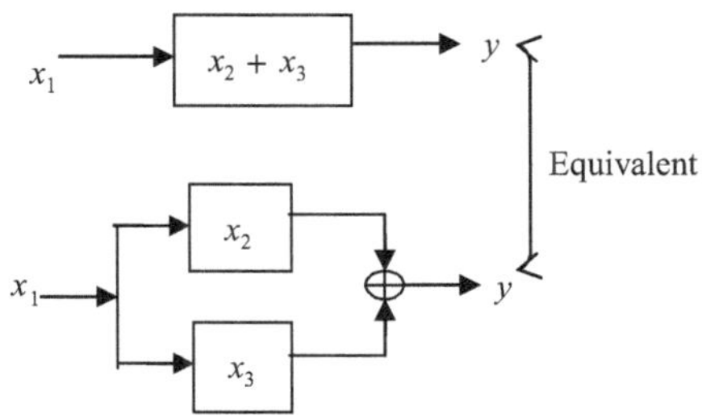

Fig. B.5 Distributive property of convolution

B.1.1.4 Transformed Domain Simplicity

Convolution in time domain is actually dual to multiplication in frequency domain. Say, given two signals are $x_1(t)$ and $x_2(t)$. The convolution sum between these two sequences is $x_3(t) = x_1(t)^* x_2(t)$.

The duality theorem states that, if the Fourier Transforms of $x_1(t)$, $x_2(t)$ and $x_3(t)$ are $X_1(\omega)$, $X_2(\omega)$ and $X_3(\omega)$. and $x_3(t) = x_1(t)^* x_2(t)$, then

$$X_3(\omega) = X_1(\omega) \times X_2(\omega)$$

Proof Referring Eq. (B.6), we can also express convolution in continuous domain as

$$x_3(t) = x_1(t)^* x_2(t) = \int_{-\infty}^{\infty} x_1(\tau) x_2(t-\tau)\, d\tau$$

Now, Fourier transform of $x_3(t)$ is

$$\begin{aligned} X_3(\omega) &= \int_{-\infty}^{\infty} x_3(t)\, e^{-j\omega t} dt = \int_{-\infty}^{\infty} \left\{ \int_{-\infty}^{\infty} x_1(\tau) x_2(t-\tau)\, d\tau \right\} e^{-j\omega t} dt \\ &= \int_{-\infty}^{\infty} x_1(\tau)\, e^{-j\omega\tau} d\tau \left\{ \int_{-\infty}^{\infty} x_2(t-\tau) e^{-j\omega(t-\tau)} d(t-\tau) \right\} \\ &= X_1(\omega) \times X_2(\omega) \end{aligned} \tag{B.10}$$

We would now try to understand convolution closely with the help of some case studies as follows. In 1st and 2nd case two phenomena in digital communication, already studied in the main chapters, will be interpreted in the light of convolution. In Case 3, a different approach of multiplication of large numbers will be discussed with the help of convolution again.

B.1.2 Case 1: Periodicity of Sampled Spectra

From the previous discussion of sampling theory of Chap. 2, we noticed that, after sampling, the spectrum (frequency response) of the analog signal becomes periodic. According to Fig. B.6, $M(\omega)$ is the spectrum of input signal to the sampler. After sampling by an impulse train, we again observe the spectrum $G(\omega)$ which is nothing but repeated $M(\omega)$ with frequency interval f_S. This phenomenon is described and interpreted using the tool of mathematics. Here, we can also justify the periodicity of sampled spectra using convolution [2].

Actually $M(\omega)$ is the spectrum of input analog signal represented using *negative frequency concept*. Therefore, the B Hz bandwidth signal is also spread in the negative axis of frequency. Now, when the signal $m(t)$ is multiplied (in time domain) with the unit impulse train $\delta_{nT_S}(t)$, the spectrum must be convolved to get the resultant spectrum. This is for; the operation convolution in time domain is equivalent to multiplication in frequency domain and vice versa. Now, if the spectrum $M(\omega)$ is taken as x (input) and spectrum[1] of $\delta_{nT_S}(t)$, (i.e., $\Delta_{nT_S}(\omega)$) is taken as h(system), after convolution, we get the resultant spectrum $G(\omega) = M(\omega)^* \Delta_{nT_S}(\omega)$.

In the CD added as *extras* with this appendix, we can closely understand the concept. There, in a presentation, the shifting of spectrum $M(\omega)$ is shown followed

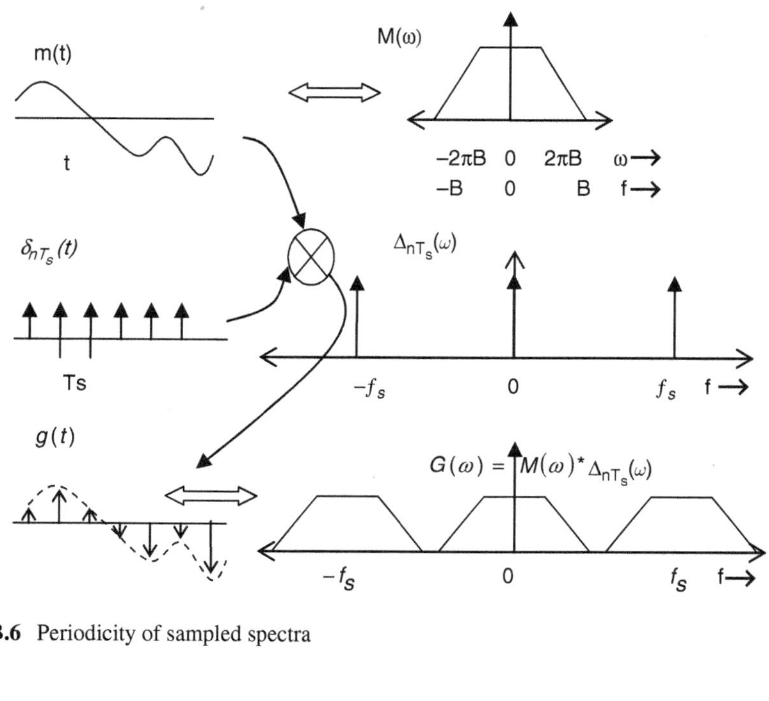

Fig. B.6 Periodicity of sampled spectra

[1] It is line spectrum, as $\Delta_{nT_S}(t)$is periodic, so always line spectrum is obtained.

by multiplication and adding. Here folding is not required as $M(\omega)$ is symmetric with respect to the Y axis.

B.1.3 Case 2: Transmission of Normally Distributed Information

From the study of information sources in information theory chapter we have understood about signal transmission, transmission probability and the concept of probability density function (PDF) of the transmitted messages. If the PDF is uniform (Appendix A), the representation is a box function or rectangular function as shown in Fig. B.7. It means the distribution is perfectly random.

Now, if more than one number of message sources are transmitting signals at a time we know the probability density function (PDF) becomes *Gaussian.* The support to the phenomena is obtained from the theory of convolution. Convolving two *rect* functions of same height and width, we get a triangular function. Moreover if one continues transmitting information, the result of convolution becomes Gaussian with varying mean,[2] standard deviation[3] and kurtosis.[4] The results are shown in the figure below and the MATLAB program is given in *http://extras.springer.com.*

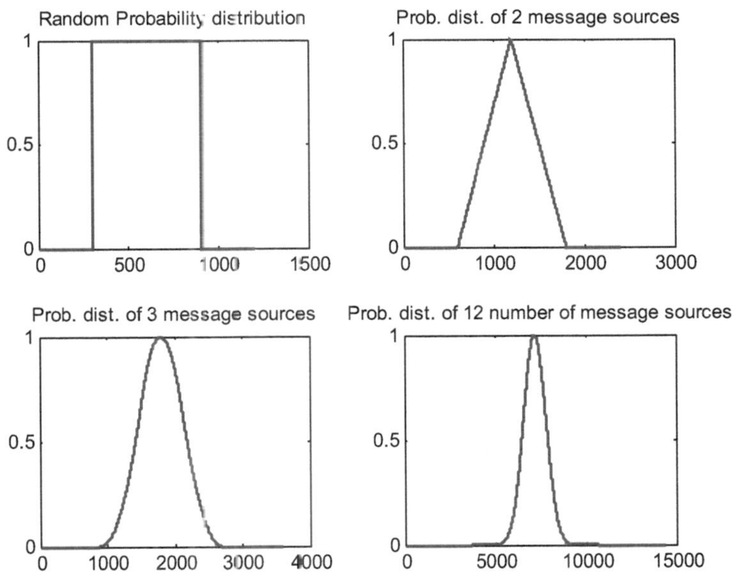

Fig. B.7 Probability distribution of randomly distributed information sources

---

[2] 1st order moment.

[3] 2nd order moment.

[4] 4th order moment.

B.1.4 Case 3: Long Multiplication Using Convolution

Convolution is equivalent to multiplication of two polynomials.

Let p1=a(0) + a(1)x +a(2)x2 and p2= b(0) + b(1)x be two polynomials in x. Then, p1p2 is given by

p1p2 = a(0)b(0)+(a(1)b(0)+a(0)b(1))x+(a(2)b(0)+a(1)b(1))x2+a(2)b(0)x3

Let a=[a(0), a(1), a(2)] and b=[b(0), b(1)]. Then convolution of a and b is given by

$$a^*b = [a(0)b(0), (a(1)b(0) + a(0)b(1)), (a(2)b(0) + a(1)b(1)), a(2)b(0)]$$

Which are the coefficients of the product of p1p2.

Let x=[3,2,4,7], y=[5,3,6,1,4] be any two arrays represented for numbers 3247 and 53614. The convolution of x and y is [15, 19, 44, 62, 59, 54, 23, 28]. The result of multiplication of the two numbers 3247 and 53614 is 174084658, which can be obtained from the result (array) of convolution.

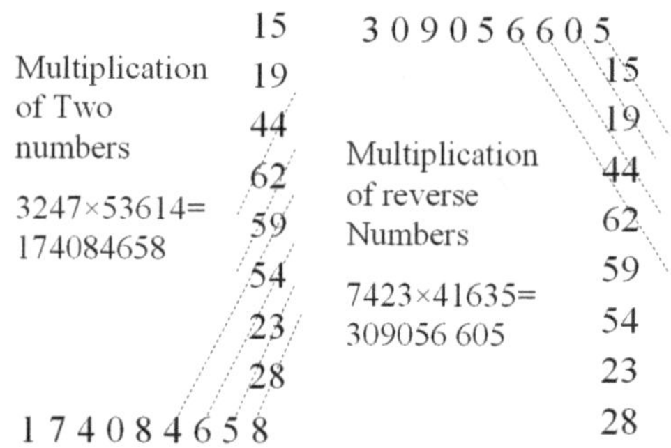

Fig. B.8 Long multiplication using convolution

Suppose the two numbers are represented in the form of arrays as x=[32,47] and y=[5,36, 14]. The result of convolution is [160, 1387, 2140, 658]. It is based 100 system of representation. The procedure of computation is as follows. Starting from the right side, keep the remainder of the 100 in the present term and add the quotient if any to the next term on the left. Repeat the same process for the terms on the left. The codes are there in http://extras.springer.com. One can go through the codes for better understanding.

B.2 Correlation

Correlation is the measure of similarity. It determines how much alike two signals are, using convolution. Correlation by itself generates a single number, called a *correlation coefficient*. A large, positive correlation coefficient indicates a relation between the signals, while zero correlation coefficient signifies that the two signals are not at all similar. A negative number indicates a negative correlation, meaning that one signal tends to decrease as the other increases.

Correlation between two different sequences is the cross-correlation and correlation between two identical sequences is the auto-correlation. Correlation can be viewed as convolution, by reversing one signal. Instead of considering signal y[k] being indexed from 0 to some positive limit K, we will *fold* it around the 0th sample, and consider that it starts at −K. For example, if y[n] = [5, 3, 1, 6], we interpret this as y[0] = 5; y[1] = 3; y[2] = 1, and y[3] = 6. If we fold this signal around, we would have y[−3] = 6, y[−2] = 1, y[−1] = 3, and y[0] = 5. When we convolve y with x, we would sum$x[n] \times y[n-k]$. Using the folded version of y, the only thing to change is the index, e.g., $x[n] \times y[n-(-k)]$ or simply $x[n] \times y[n+k]$.

As the number of data samples affects the cross-correlation measure, so the cross-correlation value must be divided by the number of data samples.

$$S_{xy}[k] = \frac{1}{N}\sum_{n=0}^{N-1} x[n]y[n+k] \tag{B.11}$$

The function below finds the cross-correlation between two signals. The total program is given in http://extras.springer.com.

correlate_apu.m

```
% Measure of similarity between two signals.
%
% Usage: [rho] = correlate_apu(k, x, y);
% inputs: k = shift (integer between 0 and length(y)-1)
% x, y = the signals to correlate
% output: rho = the correlation coefficient -1..0..1
%
% Note: This function assumes length(x) = length(y).
% Apurba Das, CDAC Kolkata

function [rho] = correlate_apu(k, x, orig_y)
% Shift y by k units
y = [orig_y(k+1:length(orig_y)), orig_y(1:k)];
N = length(x);
sxx = x*x.' - sum(x)*sum(x)/N;
syy = y*y.' - sum(y)*sum(y)/N;
sxy = x*y.' - sum(x)*sum(y)/N;
rho = sxy / sqrt(sxx*syy);
```

B.2.1 Case Study: Pattern (Shape Feature) Matching Between Two Objects Using Cross-Correlation

Here the objective is to match two different objects using the shape (boundary) feature only [1]. The matching must be invariant of scale, translation and rotation. Firstly we have taken two rectangles of different size and orientation as shown in Fig. B.9a. The distance of boundary from centroid (in rectangle that is the cross point of diagonals) of each object is taken and put into one array. This array can now behave as a one dimensional sequence. The sequences are plotted (Fig. B.9b) for different rectangles of different scale and orientation. Then the cross correlation coefficient is calculated. We found form our program that, for two rectangles the cross-correlation coefficient is 0.9919, i.e., very close to 1.

Next we have taken two objects. One is rectangle and the other is a triangle. The same method is being followed. We have again found that, the cross-correlation coefficient is 0.4232.

Therefore, we can say, the two dimensional objects can be converted to 1 dimensional sequence and then the correlation measure can infer about the respective object shape.

The complete program [3] is given using the user given input by keyboard and mouse in *http://extras.springer.com.*

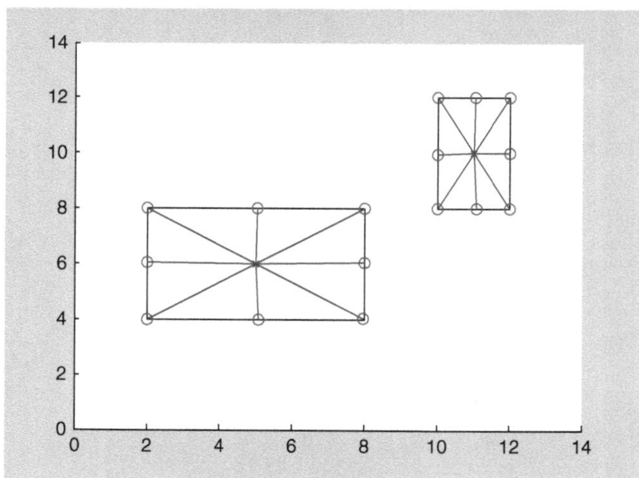

(a)

Fig. B.9 (continued)

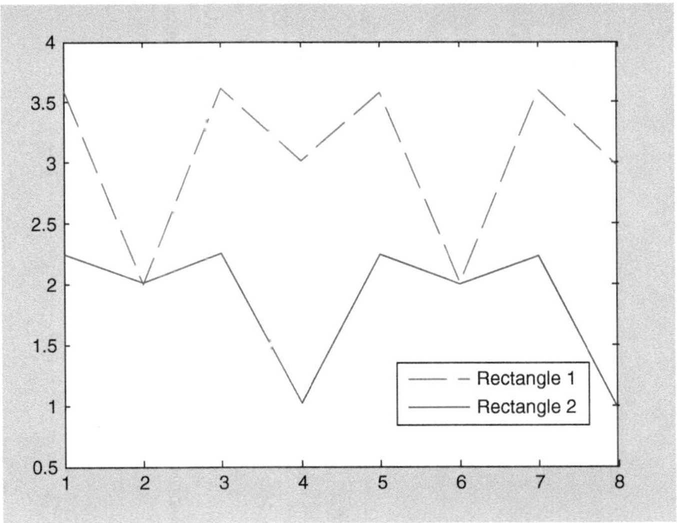

(b)

Fig. B.9 Shape matching between two rectangular objects using cross-correlation

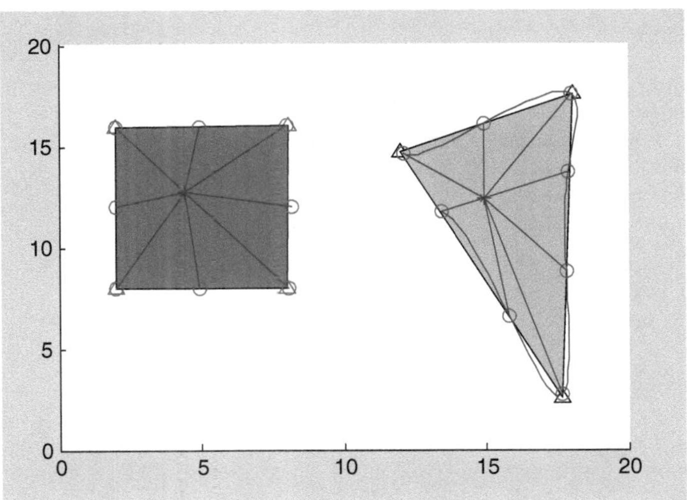

(a)

Fig. B.10 (continued)

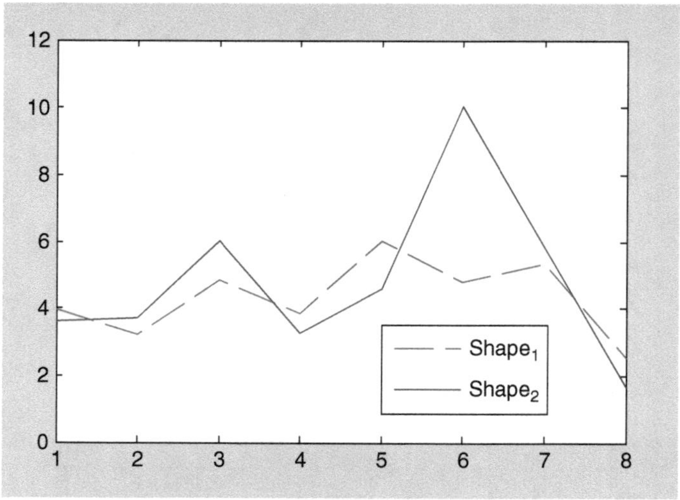

(b)

Fig. B.10 Shape matching between a rectangle and a triangle

References

1. Weeks, M., "Digital Signal Processing using MATLAB and Wavelets", Infinity Science Press LLC, 2007
2. Das, A., "Line Communication System – Telecommunication Switching Approach", New Age International Publishers, 2010
3. http://www.mathworks.com

Appendix C
Frequently Used MATLAB Functions

plot()

```
Linear 2-D plot
Syntax
plot(Y)
plot(X1,Y1,...)
plot(X1,Y1,LineSpec,...)
plot(...,'PropertyName',PropertyValue,...)
plot(axes_handle,...)
h = plot(...)
hlines = plot('v6',...)
```

Description

plot(Y) plots the columns of Y versus their index if Y is a real number. If Y is complex, plot(Y) is equivalent to plot(real(Y),imag(Y)). In all other uses of plot, the imaginary component is ignored.

plot(X1,Y1,...) plots all lines defined by Xn versus Yn pairs. If only Xn or Yn is a matrix, the vector is plotted versus the rows or columns of the matrix, depending on whether the vector's row or column dimension matches the matrix.

plot(X1,Y1,LineSpec,...) plots all lines defined by the Xn,Yn,LineSpec triples, where LineSpec is a line specification that determines line type, marker symbol, and color of the plotted lines. You can mix Xn,Yn,LineSpec triples with Xn,Yn pairs: plot(X1,Y1,X2,Y2,LineSpec,X3,Y3)

imshow()

```
Display image
Syntax
```

```
imshow
imshow(I)
imshow(I,[low high])
imshow(RGB)
imshow(BW)
imshow(X,map)
imshow(filename)
himage = imshow(...)
imshow(...,param1,val1,param2,val2)
```

Description

imshow displays the file chooser dialog box so you can select an image file interactively.

imshow(I) displays the intensity image I.

imshow(I,[low high]) displays I as a grayscale intensity image, specifying the display range for I. The value low (and any value less than low) displays as black; the value high (and any value greater than high) displays as white. Values in between are displayed as intermediate shades of gray, using the default number of gray levels. If you use an empty matrix ([]) for [low high], imshow uses [min(I(:)) max(I(:))]; that is, the minimum value in I is displayed as black, and the maximum value is displayed as white.

imshow(RGB) displays the truecolor image RGB.

imshow(BW) displays the binary image BW. imshow displays pixels with the value 0 (zero) as black and pixels with the value 1 as white.

imshow(X,map) displays the indexed image X with the colormap map. A color map matrix may have any number of rows, but it must have exactly 3 columns. Each row is interpreted as a color, with the first element specifying the intensity of red light, the second green, and the third blue. Color intensity can be specified on the interval 0.0 to 1.0.

imshow(filename) displays the image stored in the graphics file filename. The file must contain an image that can be read by imread or dicomread. imshow calls imread or dicomread to read the image from the file, but does not store the image data in the MATLAB workspace. If the file contains multiple images, the first one will be displayed. The file must be in the current directory or on the MATLAB path.

himage = imshow(...) returns the handle to the image object created by imshow

drawnow()

Complete pending drawing events

```
Syntax
drawnow
```

Description

drawnow flushes the event queue and updates the figure window.

Remarks

Other events that cause MATLAB to flush the event queue and draw the figure windows include Returning to the MATLAB prompt A pause statement A waitforbuttonpress statement A waitfor statement A getframe statement A figure statement

Examples

```
Executing the statements x = -pi:pi/20:pi;
plot(x,cos(x))
drawnow
title('A Short Title')
grid on
```

as an M-file updates the current figure after executing the drawnow function and after executing the final statement.

stairs()

Stairstep plots display data as the leading edges of a constant interval (i.e., zero-order hold state). This type of plot holds the data at a constant y-value for all values between x(i) and x(i+1), where i is the index into the x data. This type of plot is useful for drawing time-history plots of digitally sampled data systems.
Example – Stairstep Plot of a Function

```
Define a function f that varies over time, alpha = 0.01;
beta = 0.5;
t = 0:10;
f = exp(-alpha*t).*sin(beta*t);
```

```
Use stairs to display the function as a stairstep plot and a
linearly interpolated function. stairs(t,f)
hold on
plot(t,f,'--*')
hold off
```

int2str()

```
Integer to string conversion
Syntax
str = int2str(N)
```

Description

str = int2str(N) converts an integer to a string with integer format. The input N can be a single integer or a vector or matrix of integers. Noninteger inputs are rounded before conversion.

Examples

```
int2str(2+3) is the string '5'.
One way to label a plot is title(['case number ' int2str(n)])
```

For matrix or vector inputs, int2str returns a string matrix: int2str(eye(3))

```
ans =

1   0   0
0   1   0
0   0   1
```

conv()

Cell arrays can replace comma-separated lists of MATLAB variables in Function input lists Function output lists Display operations Array constructions (square brackets and curly braces)

If you use the colon to index multiple cells in conjunction with the curly brace notation, MATLAB treats the contents of each cell as a separate variable. For example, assume you have a cell array T where each cell contains a separate vector. The expression T{1:5} is equivalent to a comma-separated list of the vectors in the first five cells of T.

Consider the cell array C: C(1) = {[1 2 3]};

C(2) = {[1 0 1]};
C(3) = {1:10};
C(4) = {[9 8 7]};
C(5) = {3};

To convolve the vectors in C(1) and C(2) using conv, d = conv(C{1:2})

```
d =
        1  2  4  2  3
```

ginput()

Interactive Plotting

The ginput function enables you to use the mouse or the arrow keys to select points to plot. ginput returns the coordinates of the pointer's position, either the current position or the position when a mouse button or key is pressed. See the ginput function for more information.

Example – Selecting Plotting Points from the Screen

This example illustrates the use of ginput with the spline function to create a curve by interpolating in two dimensions.

First, select a sequence of points, [x,y], in the plane with ginput. Then pass two one-dimensional splines through the points, evaluating them with a spacing one-tenth of the original spacing. axis([0 10 0 10])

```
hold on
% Initially, the list of points is empty.
xy = [];
n = 0;
% Loop, picking up the points.
disp('Left mouse button picks points.')
disp('Right mouse button picks last point.')
but = 1;
while but == 1
  [xi,yi,but] = ginput(1);
  plot(xi,yi,'ro')
  n = n+1;
  xy(:,n) = [xi;yi];
end
% Interpolate with a spline curve and finer spacing.
t = 1:n;
ts = 1: 0.1: n;
xys = spline(t,xy,ts);
% Plot the interpolated curve.
plot(xys(1,:),xys(2,:),'b-');
hold off
```

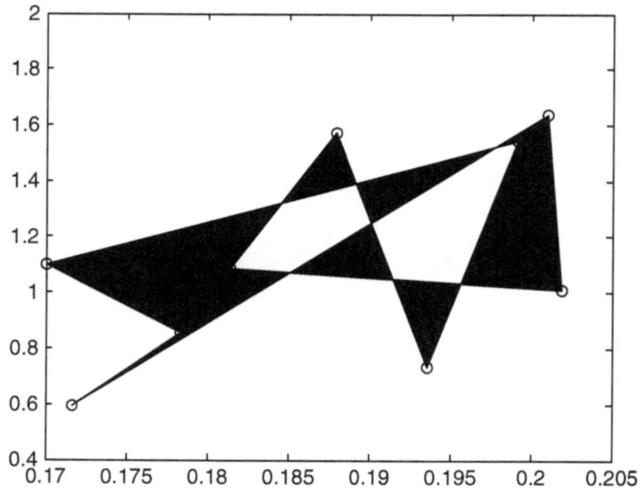

Fig. C.1 Example output of ginput()

spline()

```
Cubic spline data interpolation
Syntax
pp = spline(x,Y)
yy = spline(x,Y,xx)
```

Description

pp = spline(x,Y) returns the piecewise polynomial form of the cubic spline interpolant for later use with ppval and the spline utility unmkpp. x must be a vector. Y can be a scalar, a vector, or an array of any dimension, subject to the following conditions: If Y is a scalar or vector, it must have the same length as x. A scalar value for x or Y is expanded to have the same length as the other. See Exceptions (1) for an exception to this rule, in which the not-a-knot end conditions are used. If Y is an array that is not a vector, the size of Y must have the form [d1,d2,...,dk,n], where n is the length of x. The interpolation is performed for each d1-by-d2-by-...-dk value in Y. See Exceptions (2) for an exception to this rule.

yy = spline(x,Y,xx) is the same as yy = ppval(spline(x,Y),xx), thus providing, in yy, the values of the interpolant at xx. xx can be a scalar, a vector, or a multidimensional array.The sizes of xx and yy are related as follows: If Y is a scalar or vector, yy has the same size as xx. If Y is an array that is not a vector, If xx is a scalar or vector, size(yy) equals [d1, d2, ..., dk, length(xx)]. If xx is an array of size [m1,m2,...,mj], size(yy) equals [d1,d2,...,dk,m1,m2,...,mj].

Exceptions

If Y is a vector that contains two more values than x has entries, the first and last value in Y are used as the endslopes for the cubic spline. If Y is a vector, this means f(x) = Y(2:end-1) df(min(x)) = Y(1) df(max(x)) = Y(end) If Y is a matrix or an N-dimensional array with size(Y,N) equal to length(x)+2, the following hold: f(x(j)) matches the value Y(:,...,:,j+1) for j=1:length(x) Df(min(x)) matches Y(:,:,...:,1) Df(max(x)) matches Y(:,:,...:,end)

Note You can also perform spline interpolation using the interp1 function with the command interp1(x,y,xx,'spline'). Note that while spline performs interpolation on rows of an input matrix, interp1 performs interpolation on columns of an input matrix.

Example

This generates a sine curve, then samples the spline over a finer mesh.

```
x = 0:10;
y = sin(x);
xx = 0:.25:10;
yy = spline(x,y,xx);
plot(x,y,'o',xx,yy)
```

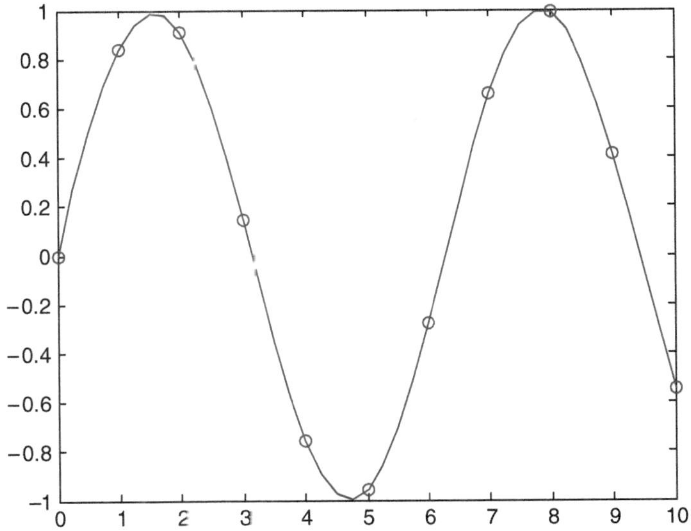

Fig. C.2 Example output of spline()

Reference

1. http://www.mathworks.com

Index

FSC
www.fsc.org
MIX
Papier | Fördert gute Waldnutzung
FSC® C083411

Kolibri 360 GmbH
Ferdinand-Jühlke-Straße 7
99095 Erfurt, Deutschland
produktsicherheit@kolibri360.de